THIEVES DECEIVERS AND KILLERS

WILLIAM AGOSTA

THIEVES DECEIVERS AND KILLERS

TALES OF CHEMISTRY IN NATURE

PRINCETON UNIVERSITY PRESS   PRINCETON AND OXFORD

Copyright © 2001 by Princeton University Press
Published by Princeton University Press, 41 William Street, Princeton,
New Jersey 08540
In the United Kingdom: Princeton University Press, 3 Market Place, Woodstock,
Oxfordshire OX20 1SY

Library of Congress Cataloging-in-Publication Data

Agosta, William C.
Thieves, deceivers, and killers : tales of chemistry in nature / William Agosta.
    p.   cm.
Includes bibliographical references (p.).
ISBN 0-691-00488-9 (alk. paper)
1.Chemical ecology.   I. Title.
QH541.15.C44 A49 2000
577—dc21   00-032627

This book has been composed in Palatino

The paper used in this publication meets the minimum requirements of
ANSI/NISO Z39.48-1992 (R1997) (*Permanence of Paper*)

www.pup.princeton.edu

Printed in the United States of America

10 9 8 7 6 5 4 3 2 1

CONTENTS

THIEVES DECEIVERS AND KILLERS

PROLOGUE

The Protos and Their Slaves

The three Proto scouts moved closer. They saw what looked like a Lept camp up ahead but wanted to make sure. Soon they were close enough to spot the entrance and, sure enough, there the Lepts were, going in and out, carrying food. The Protos had been right: It was a medium-sized Lept camp and near enough to their own for a quick attack. This must be the camp of the Lepts they had been encountering in recent days. Two of the Proto scouts turned back toward home to gather reinforcements, while the third stayed behind to guide the raiding party to the camp entrance.

The two scouts made good time returning home and announced their news at once. Excitement ran through the Proto compound as the story spread. A Lept camp only a short distance away! For as long as anyone could remember, the Protos had kept Lepts as slaves, and like slave-owners everywhere they needed to renew their supply of captives from time to time. For the Protos, nothing really worked well without Lepts. Without an ample supply of slaves, their whole society would disintegrate. It was not merely that the Protos were lazy, but they had kept slaves for so long that they were truly unable to care for themselves. The slaves could raise young, gather food, and keep the place clean, whereas the Protos themselves really excelled only at capturing Lepts. Otherwise, they spent a great deal of time lolling around their camp, acting bored, and asking the Lepts for something to eat. Lept slaves did all the work and were remarkably faithful to their Proto masters, even becoming ferocious participants in slave-raids on their own kind.

As the scouts made their report, a well-armed raiding party of both Protos and slaves gathered around them. When the group was large enough, the scouts led it out toward the Lept camp. One scout led the way over the carefully marked trail, the eager party briskly following in an orderly column. They soon made their way to the Proto scout left behind earlier and grouped excitedly for the assault.

Intimidating the few Lepts encountered on the way, the raiders advanced directly to the entrance. They forced their way through the entrance, attacking and pushing their way into the camp. The Lepts did not give up easily, and soon the fighting was fierce. As the battle proceeded, the distinctive fighting styles of master and slave became apparent. The Lept slaves fought hard, slashing and stabbing the free Lepts who were defending their home. As Lept struggled against Lept, combatants were equally matched, and the clashes often ended with one or both participants injured or dead. Encounters between Lepts and Protos, however, were generally quite different. Though the Lepts were brave, they rarely attacked the Protos. The Protos themselves almost never struck a blow. If a Lept did strike a Proto, the fast-moving Proto slipped neatly away. Instead of direct physical combat, the Protos preferred using an impressive chemical weapon that each one carried. The Protos needed to spray only a few drops of this chemical and the Lepts were reduced to panic and confusion. They forgot about challenging the invaders but turned instead on one another, suddenly fighting among themselves in chaos and turmoil. All Protos carried this unusual chemical spray but did not share it with their slaves. Although the slaves fought fiercely, the Protos' weapon really won the battle. In fact, the Protos generally had no trouble defeating Lepts even when there were no slaves in the raiding party.

Meanwhile, as the conflict continued, several raiders pushed deep into the camp, searching out the Lepts' communal nursery. Here, where the Lepts kept their young, was the Protos' real goal. Because adult Lepts would never adapt to enslavement, the Protos had no interest in them. A Lept captured at birth, however, and raised among the Protos would accept servitude without question. Thus the raiders began removing infants from the nursery for the trek back to the Proto compound. Lept nurses rushed about trying to save their wards, grabbing up as many as they could and running for the camp entrance and the safety beyond. Many reached the entrance, but two Protos stationed just inside barred their way to freedom. These guards did not harm the nurses but allowed them to escape only after they released their precious loads. In this way, the young Lepts were captured and the adults were either destroyed or cleared from their camp and dispersed.

After the battle, several Protos carried the little ones back to the Protos' camp, making repeated trips until the Lept nursery was emptied. They installed the captured infants in the Proto nursery where, under the eye of slave nurses, the little Lepts would learn their place in the world. This Proto home would be the only one they would ever know, and they would grow to adulthood unaware of what might have been. Any free Lept they met they would recognize only as an enemy.

For the Protos, it had been another successful raid. Several slaves had died, but the Protos themselves suffered few injuries. Owing to their chemical weapon, the Protos could raid the Lepts with little loss of life to themselves or their victims. If any society based on slavery could be characterized as "advanced," the Protos certainly seemed so. Other slave-raiders that preyed on Lept nurseries killed as many adults as possible during their assaults. Some even ate their victims. In comparison, the Protos could pass as relatively civilized. From the Proto point of view, of course, charity made good sense. By letting the adult Lepts escape, the Protos assured themselves of future Lept nurseries to plunder. The day after the raid, in fact, the scattered Lepts straggled home to pick up their lives and begin replenishing their looted nursery.

From this portrayal, we might condemn the Protos as barbarous creatures, but as you may have known all along the Protos are not wicked slave-keeping humans. Protos and Lepts are tiny ants that live out their entire lives in a world no larger than a dinner table. A Proto camp consists of one or two dozen indolent masters cared for by perhaps twice as many Lept slaves, all sheltered in an empty milkweed stem or perhaps within a hollow acorn. Free Lepts form similar colonies of two or three dozen individuals. The Protos' formal name is *Protomognathus americanus,* and the Lepts they enslave are *Leptothorax curvispinosus* and two other closely related species. Their little domains occur throughout the broadleaved woodlands of eastern North America, from Ontario to Virginia and as far west as Ohio.

Slave-making ants' remarkable habits have made them a favorite of natural scientists for almost two hundred years. Generations of entomologists (scientists who study insects) have delighted in watching them in the field and arranging their raids in the laboratory. If the ants' behavior seems almost human, scientists have also noted a contrary aspect that we should not overlook. Whereas sight and sound are human beings' primary modes of communicating with one another, many of the ants' interactions depend on releasing and sensing chemicals.

If Lepts and Protos were people, they would identify friend and foe by sight. They would mark their trails with visible signs and exchange information using written and spoken language. Instead of depending on sight and sound, however, Lepts and Protos communicate with chemicals. Other kinds of ants do the same. On meeting, two ants touch each other with their antennae, "smelling" the chemicals on each other's body for identification. They readily distinguish other ant species from their own, and within their own species they discriminate members of their home colony from outsiders. To mark their trails, scout ants deposit minute amounts of chemicals on the ground behind them as they move along. The Proto scouts that returned home to report finding the Lept camp used chemicals in communicating their discovery. A chemical recruitment signal they released helped assemble the raiding party. The Protos probably identified and located the Lept brood by odor as well. Overall, ten to twenty different antenna-detected chemical signals are necessary to assure an ant colony's smooth operation.

Chemicals also have a critical place in ant warfare. Most ants possess a sting that delivers poisonous chemicals to their enemies. In the strenuous fighting of the slave raid, Lept slaves and free Lepts frequently stung one another severely. Some kinds of ants employ other chemical warfare agents as well, such as the Protos' unusual weapon that created turmoil among the Lepts and caused them to attack one another rather than the invading Protos.

It was ants' extensive dependence on chemicals that brought Lepts and Protos to the opening of our story. These little insects vividly illustrate chemical interactions in creatures' lives. Just as we considered Protos and Lepts, we want to look at other living creatures as they employ their chemicals, sending messages, defending themselves, and carrying out many other activities.

The world of chemical relations among plants, animals, and other organisms was completely hidden from us until only a few decades ago. Many scientists thought that chemicals now recognized as central to these relations were nothing more than cellular waste products. Interactions among organisms were much less well understood in the past, and the role of chemicals in many of them was not yet acknowledged. Biologists and chemists opened up this world of interactions among living creatures only as appropriate chemical and biological investigative tools started to appear around mid-century. Only then did the story of these universal chemical activities begin to take shape.

As the story of the Protos' raid demonstrates, a familiarity with chemistry or biology is certainly not required to delve into these activities. An up-close look at the chemicals used by the Protos provides a marvelous entry into a wider biological world filled with events as remarkable and fascinating as the raids of slave-making ants. Other creatures have equally amazing stories and, from a broader point of view, their behaviors play a key role in the health of the earth, the preservation of environments and biodiversity, and the extinction of species.

The chemicals featured in the behavior of many organisms also touch our own lives in important ways. They provide a sizable fraction of modern medicines, as well as perfumes, pesticides, and other products ranging from textiles to glue. Some of these chemicals have been in use for thousands of years and have intriguing histories. Others offer ways to save threatened environments; all affect our own lives, and some do so profoundly.

Most of us now live far from the close daily contact with the natural world that was taken for granted until well into the nineteenth century. It is easy to forget that, as living creatures, we too are part of this natural world and that our high-tech existence today depends on this world and its creatures no less than life did in former times. In what follows, we shall see what the chemicals of this natural world are all about.

From Protos and Lepts to Nature's Special Chemicals

1

The Protos' warfare on their Lept neighbors depended heavily on chemicals, but ants are by no means unique in making extensive use of chemicals for communication and warfare. From one-celled organisms to complex plants and animals, many living creatures do the same. As species develop over evolutionary time, it is relatively easy for them to adapt their cellular machinery to producing chemicals for communication, warfare, and other purposes. These chemicals facilitate the way of life of organisms spread all across the biological spectrum.

Like the signals of the Proto scouts, one large group of chemicals carries messages that pass between members of the same species, messages that humans can express in words. Such signals are conveyed by substances called pheromones, which can transmit different kinds of information. Some pheromones are attractants, bringing male and female together for mating, or perhaps assembling a group of creatures for feeding or defense. Others carry such varied messages as "Danger! Flee!" "This is my territory," or "I am pregnant."

Other types of chemical signals that occur may pass between members of different species. These interspecific signals, as they are called, may benefit one or both of the species involved. The Protos'

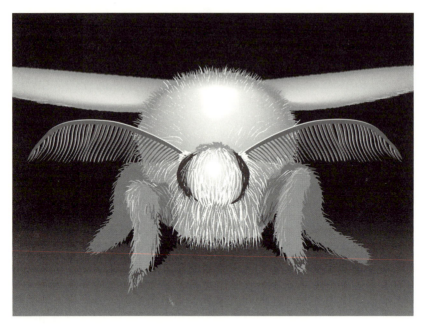

ILLUSTRATION 1 About half the 15,000
to 20,000 hairs on this male silkworm moth's
feathery antennae are specialized for
detection of the female's sex attractant
pheromone.

chemical warfare weapon, for example, can be viewed as an interspecific signal that benefits only the sender as it reduces the recipi-
ent Lepts to fighting among themselves. On the other hand, sometimes the receiver is the sole beneficiary of a message, as when a
predator locates its prey by following the distinctive odor the victim haplessly communicates to its enemy. Other signals between
species serve both sender and receiver, as when a flower's delight-
ful fragrance entices a foraging insect to linger and explore. In an
exchange profitable to both organisms, the insect pollinates the
flower and receives a drop of nectar in return.

Chemicals may also supply more general information about a
creature's environment. Salmon return from the sea to their native
stream to spawn, guided by the distinctive odor of their birthplace.
Thirsty animals follow their noses to locate life-saving water. Other

chemicals are closely associated with organisms' characteristic ways of life, such as the silk spun into a silkworm's protective co-coon or the pearl an oyster fashions around a grain of sand in re-sponse to the gritty irritation.

Organisms, wherever they are, interact with one another to live, and chemicals mediate many of these contacts. Some inter-actions may be optional and occasional, whereas others are abso-lutely necessary to sustain the organism's way of life. Protos can survive only by keeping Lepts as slaves. Various species of slave-making ants differ in their self-sufficiency, but Protos have lost the ability to care for themselves. Deprived of its slaves, a Proto colony soon deteriorates.

Before continuing with the role of chemicals in the living world, we should clarify certain words and concepts that we have been using. The chemicals we have been talking about are chemical com-pounds made of molecules that are three-dimensional arrange-ments of atoms joined together by chemical bonds. A molecule is the smallest existing unit of a chemical compound, just as an atom is the smallest existing unit of a chemical element. The specific ar-rangement of the atoms in a molecule—how the atoms are joined together—is called its molecular structure. It is this structure that makes a molecule what it is. Molecules come in a multitude of sizes, from as few as two atoms to many thousands. The atoms themselves are of many types, but the biologically active molecules that interest us here are usually limited to atoms of carbon, hydro-gen, oxygen, perhaps nitrogen, and, less frequently, a bit of sulfur or phosphorus. Many of these biomolecules are medium sized, with twenty to eighty or ninety atoms. This size range also includes the molecules of such familiar natural compounds as cocaine, peni-cillin, and cholesterol.

Chemists can make, or synthesize, many biologically active chemical compounds in the laboratory. (Chemical, chemical com-pound, and compound are interchangeable terms.) One starts with readily available simple molecules and proceeds to build more complex structures, one step at a time, through chemical reactions.

This procedure is called stepwise synthesis, and chemists have used it to prepare thousands of compounds that are found in nature, as well as many thousands more that have never been found in living systems. The chemist's repertory comprises many different synthetic reactions that can be carried out in different combinations, so that numerous ways are often available for preparing a specific compound. Living organisms, too, make molecules by stepwise syntheses, employing their enzymes (natural proteins that speed up, or catalyze, chemical reactions) and biochemical machinery rather than the chemist's laboratory reactions. Typically, nothing about an ordinary biomolecule reveals its history or origin, so a particular compound synthesized in the laboratory is indistinguishable from the same one obtained from nature.

Chemists can also study a novel compound from nature, find the specific connections between its atoms, and thus determine its structure. For complex molecules this can be a difficult research problem, particularly if very little of the compound is available. Over the past twenty-five years, however, much structure determination has become routine. Chemists have identified the molecular structures of some of the chemical compounds we shall encounter, but the structural details need not be part of our story. Nonetheless, it is important to keep in mind that molecules are real physical objects; depending on structural details, they may be large, small, globular, flat, floppy, or rigid, but all have mass and occupy space. It is also worth emphasizing that a compound's molecular structure is the basis of its unique physical, chemical, and biological properties. The chemical compounds of living organisms often have remarkable characteristics that proceed directly from the compounds' structures, sometimes in ways that we do not yet fully comprehend. Learning the relationship between molecular structures and the properties they imply remains one of the most intriguing scientific problems.

The chemical compounds that will interest us generally are significantly different from the three or four dozen essential compounds required to sustain life. Some of these essential compounds are building blocks for creating the proteins and nucleic acids (such as DNA) that are necessary for life as we know it. Whether proteins come from blue mussels, fir trees, or hippopotamuses, they consist of the same twenty amino acids strung together, one after the other, in different sequences. Similarly, each organism's DNA incorporates four units, known as nucleotides, that are used over and over again but are strung together in variable order. These amino acids and nucleotides, along with approximately two dozen other compounds, are indispensable components of life on earth.

The compounds that we will consider here, in contrast, serve quite specific needs of various organisms, and for this reason we will call them *special* chemical compounds. Different species may employ the same special compound, in some cases for the same purpose and in other cases for different ends. For example, the carbon dioxide arising from the respiration of a crowd of ants is an aggregation signal that invites solitary ants to join their nestmates. Corn rootworms, however, use the carbon dioxide that living corn roots emit into the soil as a signal, leading them to their food. Different species may also employ diverse compounds for essentially the same purpose. Various ant species mark their food trails with different chemicals to keep their food sources secret from one another.

As far as we know, one species, or a few closely related ones, often have a monopoly on a particular compound. The tobacco plant (*Nicotiana tabacum*), for example, synthesizes nicotine to defend itself from attack by herbivores (animals that feed on plants) and as a convenient way of storing nitrogen temporarily. Neither roses, elephants, nor starfish contain nicotine. In fact, apart from a few of tobacco's close relatives, we know of no other living organism that makes nicotine. In the same way, other plants have devised their own defensive compounds for protection from herbivores.

One may wonder why chemical defenses differ from plant to plant, or why carbon dioxide attracts certain creatures but not others. How does a special chemical and the interactions associated with it come to have a place in the life of a particular species? Although we do not have detailed histories of most special chemicals, we can answer in general terms that, like other biological adaptations and developments, they have arisen through the process of biological evolution. Because this topic comes up more than once in our story, it may be useful to say something about it here. Evolution is based on straightforward concepts, but their application can be tricky and is often misunderstood.

Evolution is a two-step process that leads to changes over time in the genes (all organisms' hereditary units, composed of nucleic acids) of a population of a species. After describing the two steps and their consequences, we can discuss examples that should make the process clearer. Step one is genetic variation in every generation throughout the living world. This variation arises mainly through chance mutation of genes, which biologists can now understand on a molecular level. It is this variation that, for example, may result in an insect population that is more resistant to pesticides than its relatives are or individuals with a genetic propensity for a disease such as cystic fibrosis. Genetic variation is natural and unavoidable; without it the population of an organism could not adapt to changing environmental conditions over time.

With genetic variation, adaptation to change is possible by way of an operation called natural selection, which is step two. Owing to genetic variation, individual organisms vary in ability to deal with their environment. As a consequence, some have a greater chance of surviving and reproducing than do others. In most species only a small fraction of individuals survive. Such creatures as goldfish and oysters produce millions of offspring, nearly all of which meet an early death. Survival is mostly a matter of luck, but offspring with genetic makeups best suited to their environment have some advantage. In the long run, such genetically favored organisms are the ones most likely to reproduce, and thus pass their genes on to their progeny. As a result, over time the genes favoring survival and reproduction can spread through a population of organisms, altering the population's characteristics. In this way, the population evolves.

These abstract ideas can be concretely illustrated. The English peppered moth (*Biston betularia*) is a classic example that is frequently cited in discussions of the concept of evolution. The peppered moth's color is under the primary control of a single gene, and thus is inherited. Genetic variation (step one) has produced two forms of this gene, which cause an individual moth to be either light-colored or dark. Although the two color forms exhibit no other differences, the numbers of light and dark moths are not equal. In the early nineteenth century, nearly all the moths were the light variety and naturalists rarely collected a dark moth. The natural bias against dark moths has a simple explanation. The moths are active at night and spend the daylight hours at rest. Their resting sites were relatively light in color, rendering light-colored resting moths nearly invisible and dark ones quite prominent. As a result, birds found and ate nearly all of the dark moths, but few of the light ones. The birds are agents of natural selection (step two), which in this case disfavors dark moths.

By 1898, however, nearly all the peppered moths in the cities of the north of England were the dark variety. Decades of uncontrolled industrialization had filled the air with soot and pollutants. The ever present soot had darkened the entire countryside, and thus resting moths now lay against a dark background. Once again, hungry birds were agents of natural selection, but now they could more readily spot and consume the light moths. Under these new circumstances, the dark form prospered.

The further passage of time brought still another shift. In the mid-twentieth century, as environmental laws controlling industrial emissions came into effect, soot and pollutants disappeared from the air. Tree trunks and the moths' other resting sites took on their original lighter hue. Again disfavored, dark-colored peppered moths have now begun to disappear from the industrialized areas where they had predominated. The light form, once again

ILLUSTRATION 2 Light and dark
peppered moths. In the late nineteenth
century, the dark form had a 30 percent
better chance of survival than the light form.

better camouflaged and protected from predation, is returning to
dominance.

The reversals in the moths' dominant color are not only splendid examples of evolution, but they also illustrate important attri-
butes of the process. One is that natural selection operates on individuals, not genes. The birds devour particular moths; they know
nothing about genes. Also, it is not the individuals that evolve over
time, but the population. A single moth undergoes no evolutionary
change. It is born, lives, and dies either as a dark moth or as a light
one. Most significant, the fluctuating fate of light and dark moths
demonstrates a general characteristic of evolution: It is not "progress" in any broad sense. The favored color of moth is better strictly
with reference to the prevailing local conditions. Neither color form
is an absolute improvement over the other. Twice, when conditions
changed, a shift occurred in the favored color. Expressed in another

way, natural selection is not a guided force; it is simply an effect. Birds eat the moths they can find, and they apparently find nearly all the more visible ones, whatever their color. Comparatively few of these survive to reproduce.

For the peppered moth, genetic variation provided light- and dark-colored forms. Genetic variation is also apparent in a familiar biological development going on around us all the time. Over the past fifty years we have used antibiotics to treat infectious diseases, to poison and destroy the pathogenic bacteria that cause these afflictions. It is well known that pathogens frequently develop resistance to an antibiotic over time and, in fact, some pathogens are now resistant to most available antibiotics. This resistance begins with natural genetic variation (step one). Owing to their genetic differences, individual bacteria vary in their response to a particular antibiotic. Some tolerate it better than others of their species. The more tolerant ones have a better chance of surviving, reproducing, and passing on to their offspring the genes responsible for their greater tolerance. Here the antibiotic is the agent of natural selection (step two). Over generations of such selection, the genes conferring resistance spread through the population of a particular species. Continuing exposure to the antibiotic leads to continuing selection for more resistant individuals. Because the time between successive generations of bacteria is measured in minutes or hours, this process can be relatively rapid. In this way, an entire population of the pathogen may eventually become totally resistant, and the antibiotic will have become useless against it. Once again, it is the population that evolves, not the individuals.

Resistance to antibiotics has a close parallel that has become equally familiar over the past fifty years. On repeated exposure, agricultural pests typically become tolerant of a pesticide that is used to control them. Farmers know that the amount of pesticide required to achieve control increases from year to year. Insect pests develop resistance through evolution in the same way that bacteria do. One or two serious pests are now essentially immune to all available pesticides, and many others are moving in that direction.

These specific examples of evolution are relatively straight-forward and interpretable. However, the evolution of special chemicals and their biological interactions is more complex and may take place over millions of years rather than only a few decades. Such developments are rarely understood in detail, but despite the greater complexity evolution here also proceeds by way of repeated genetic variation and natural selection. Some striking examples appear in the chapters that follow.

Before delving into ways the living world uses its special chemicals, we should note that these compounds touch our own lives in important ways. For millennia, humans have been borrowing natural chemicals for their own purposes, most often as drugs. Our oldest medicine is opium, which we prepare from the opium poppy (*Papaver somniferum*) today much as Mediterranean peoples did four thousand years ago. Just as we do, these early communities valued opium for its ability to kill pain and impart a sense of well-being. The principal constituent responsible for these effects is a chemical compound called morphine, which remains unsurpassed in its ability to control severe pain. In poppies, morphine's toxicity and bitterness presumably repel herbivores looking for a tasty meal.

In terms of monetary value, pharmaceuticals are our most important products obtained from nature, but natural chemicals have long filled other needs as well. Closely allied to drugs are compounds that we call recreational chemicals, most of which come from plants and find general use as mild stimulants. Caffeine from coffee and tea and nicotine from tobacco are the most popular ones. In addition, there are natural products unrelated to drugs. Probably the most ancient of these is a Chinese discovery from almost five thousand years ago. The early Chinese mastered the art of converting a protein from silkworm cocoons into the splendid textile we know as silk. For several millennia, silk manufacturing has not changed fundamentally from the Chinese process originated long ago. Other ancient developments include textile dyes and hunting poisons, as well as the perfumes and incense that have enriched human society and religion since Egyptian times. For centuries, perfumers have created alluring scents by blending dozens of natural odorants, such as musky animal sex attractants and sweet-smelling floral oils.

Another natural product that deserves special mention is the naturally occurring pesticide. The possibility of controlling pests with natural, so-called biorational, preparations is a long-standing dream that probably goes back to classical times. In his herbal, *Theatrum Botanicum*, published in 1640, John Parkinson noted that books written in ink containing wormwood (*Artemisia absinthium*) are protected from the ravages of hungry mice. He credited this observation to the second-century Greek physician Galen. Parkinson also recommended common speedwell (*Veronica officinalis*) and several other plants as useful repellents for clothes moths.

Modern interest in natural pest control rests on the high price we pay for our dependence on chemical pesticides. These synthetic agents have phenomenally improved crop yields in the Western world over the past fifty years, but the agricultural miracle has cost us dearly. For decades, indiscriminate application of pesticides has led to environmental disruption on a vast scale. The pollution of rivers, lakes, and coastal waters, as well as the wholesale destruction of beneficial organisms, are now worldwide problems. Nonetheless, chemical pesticides remain a twenty-five-billion-dollar world market. According to a 1997 federal report, there has been an encouraging decline in coastal pollution (from pesticides and other causes), and environmental improvement is under way. Despite this progress, pesticide-related damage in the United States runs to hundreds of millions of dollars each year.

Natural pesticides show promise for alleviating the pollution problem. Plants make some of these pesticides for their own protection, and bacteria synthesize others for purposes that are poorly understood. In spite of their promise, however, products based on natural pesticides have been on the market for years without much success, their major drawbacks being high cost and a reputation for unreliability. The present hope is that continuing research can transform these natural agents into attractive, dependable alternatives to traditional chemical pesticides.

Another approach to replacing chemical pesticides centers on pheromones. For over twenty-five years, agricultural scientists have explored ways of controlling insect pests by means of chemical signals. Early efforts focused on using a pest's own sex attractant to draw insects to traps or to interrupt their mating. A more recent innovation employs the attractant pheromones of beneficial species that prey on pests. The idea is to lure useful insects to the garden or field, where they can devour invading pests before the pests destroy crops.

Instead of converting compounds from plants into medicines, we can often benefit simply by eating the plants themselves. In some cases, consuming a plant as food has a demonstrably better effect on health than using a single plant constituent as a medicine. We are far from understanding fully the connections between diet and health, but nutritionists now urge us to eat at least five servings of fruits and vegetables each day to remain healthy. At the same time, cancer experts estimate that 30 percent of all types of cancer are linked to diet. Typically, we do not know just what natural compounds in foods are medically important, although new information appears all the time. In early 1997, University of Illinois scientists reported that a common dietary constituent, a compound called resveratrol, helps prevent cancer. Mulberries, peanuts, and grapes are particularly good sources of resveratrol, and the compound passes from grapes into wine. This leads to the attractive possibility that wine may someday emerge as an effective anticancer agent. At about the same time as the report from Illinois, a Harvard group found that a diet rich in red tomatoes reduces the incidence of prostate cancer up to 34 percent. A parallel Italian investigation extended the benefit of tomatoes to other types of cancer as well.

The prospect of turning natural compounds to practical use has fueled much human activity for thousands of years. Where once trial and error guided the search for beneficial chemicals, now biology and chemistry set the course. Agricultural research and the quest for new drugs provide most of the supporting funds, but investigations extend beyond these immediate goals. The result is a lively interdisciplinary undertaking that attracts scientists from several backgrounds, some seeking marketable pharmaceutical or agricultural products and others pursuing more fundamental problems. In one way or another, these scientists are probing chemically mediated interactions between living organisms and their environments. Because relations between organisms and their environments are the subject matter of ecology, the interdisciplinary study of the chemically mediated interactions goes by the name of chemical ecology.

Our interest in the importance of chemicals in organisms' lives then falls into the area of chemical ecology, a subject particularly accessible to curious laypersons. Unlike some other fields of active research, chemical ecology's initial studies are often framed as straightforward and readily comprehensible questions, such as "How do Proto scouts guide a raiding party to a nest of Lepts?" Once you know the story of the Protos' raid on the Lepts, this question is a reasonable one, even if you know nothing about chemistry or biology. Such accessibility permits anyone to appreciate the sorts of questions one kind of research scientist hopes to answer. It is easy to see how the answer to one question can lead to new questions to investigate.

Despite chemical ecology's accessibility, knowledge has come slowly. Scientists are only now beginning to fit this knowledge into the broader structure of biology. Progress was slow for decades because interdisciplinary subjects, such as chemical ecology, were long neglected. Research was partitioned into traditional disciplines, with the result that, for years, chemistry and biology moved forward on separate tracks, leaving poorly explored much that lay between them. This partitioning has now largely disappeared, but the barriers were once substantial.

Progress in chemical ecology also awaited the development of suitable chemical techniques. Some plant products are readily available, but organisms make and use many ecologically important compounds in only tiny amounts, and then often mixed with closely related substances. Until about 1960, chemists simply lacked methods for identifying the components of such mixtures. Once scientists devised the proper tools, chemical ecology (and several related subjects) began to flourish. Over the years, the tools have continued to improve, and investigations have accelerated. For example, the first pheromone to be identified chemically was the sex attractant of the silkworm moth (*Bombyx mori*). Its identification in 1959 came after twenty years' effort and the dissection of hundreds of thousands of moths. Today, similar research often requires no more than several weeks' work with a few insects, largely as a result of improved tools.

As things now stand, we know something about the special chemicals of a very small fraction of the 1.4 million species of living organisms that scientists have formally described. Several major groups of creatures have received essentially no attention. Our knowledge is much more limited than even this suggests, because the formally described species represent only a very modest fraction of the total number that inhabit the earth. No one knows what this total number is, but biologists now talk seriously of tens of millions of living species, and some authorities suggest as many as one hundred million! We are only beginning to appreciate the extent of chemical interactions in the biological world and to elucidate the details of a small fraction of them.

A number of these interactions form the basis of the accounts in the following chapters. After a few final general comments, we can commence our exploration. Many of the activities we look into are clearly purposive; like the Protos' slave raids, they are directed toward a specific end. Such purposive activities may appear to demand considerable intelligence, but as a rule they are neither rational nor intentional. They result not from an organism's careful planning but represent built-in responses to particular situations and signals from the surroundings. These responses arise through the workings of evolution and render an organism more fit and better equipped to survive. Protos lay chemical trails, raid Lept colonies, and exhibit other extraordinary behaviors, but they do not lay battle plans or discuss strategy among themselves. My opinion is that this in no way detracts from their fascination, nor does it make their story any less engaging. For me, the very existence of Protos and Lepts is in itself astounding, and the same is true of many other creatures and interactions described in the chapters that follow.

In an earlier book on pheromones (*Chemical Communication: The Language of Pheromones*), I arranged the material biologically from microorganisms to humans, in a sweep of chemical signals across the biological world from simple to complex. In a broader discussion of special chemicals (*Bombardier Beetles and Fever Trees*), I organized the chemicals by function: pheromones, plant-defense chemicals, lifestyle chemicals, and so forth. In the current work, more than one kind of organism and several different signals appear in some of the accounts. The stories stand on their own and emphasize the accounts themselves rather than fit into a higher organizational structure. Some accounts are not yet complete and stop short of a full picture, although enough is known to make them worth relating now. In practice, a research scientist's investigations are like this more often than not.

Each organism's scientific name, that is, the Latin name of its genus (plural, *genera*) and species, appears when it is first mentioned, as seen with tobacco, wormwood, and several other organisms referred to above. Larger organizational groupings, such as family or phylum, occasionally appear by name, but I have kept such formalities to a minimum. Following the text, there is a glossary that includes comments on classification of organisms and units of measurement, as well as a list of suggested further reading.

Paying Ants for Transportation

2

Just as Lepts and Protos tend their eggs and larvae in nurseries, other insects also help their young get started in various ways. Hatchlings may require a special diet, and their mother often makes sure the right food is nearby when they emerge from their eggs. Solitary wasps select an appropriate caterpillar and lay their eggs deep within its body. On hatching, the wasp larvae find themselves immersed in the nourishment they need and devour the living caterpillar from within. Similarly, fruit flies lay their eggs in the plums or cherries their larvae must eat to thrive. Other parents care directly for their offspring, remaining with them and providing food as needed. Ambrosia beetles maintain fungal gardens that they harvest for their young. Dung beetles take their name from the nutritious balls of dung they roll to their underground nests to nourish their larvae. Sexton beetles are no less diligent, efficiently burying the corpses of small mammals to provide a protein-rich diet for their brood and themselves.

These parents labor industriously for their progeny, but some others seem to be casual and negligent. Among the apparently negligent parents are some odd creatures called stick insects, which belong to an order or group of insects known as Phasmida. The stick insects that interest us are slow-moving leaf-eaters, often camouflaged by thin, stick-like bodies in green, gray, or brown. Able to remain immobile for hours, they are almost invisible as they cling to a stem or twig. Some of them are quite large, and, in fact, the longest known insect is a tropical stick, *Pharnacia kirbyi*, whose female can reach a length of 36 centimeters (14 inches)! If attacked, many stick insects simply fall to the ground and play dead. The small males usually have wings and can fly about, but the females either have no wings or fly poorly, particularly when heavy with eggs. Owing to their sedentary habits and curious stick-like appearance, these quiet-living creatures have long been popular as insect pets.

To lay her eggs, the female stick insect hangs from a leaf or twig and releases one egg at a time. Many species simply allow their eggs to drop to the ground, but some females toss each one away with a quick flick of their abdomen. In either case, this is the end of the adults' involvement with the next generation. Their duty is done, and the eggs are on their own as soon as they land in the leaf litter, often concentrated in a small area beneath the female. This seems like a poorly designed operation, because eggs confined to a restricted area offer predators a more rewarding target than eggs that are more widely dispersed.

The eggs are slow to hatch, but they finally yield nymphs that are active little creatures resembling miniature adults. (This form of development is called gradual or incomplete metamorphosis, and the young are often referred to as nymphs.) The stick nymphs run about, here and there, until they come upon an upright support to climb. Their instinct is to ascend the first upright they encounter, climbing until they reach foliage to feed upon. This means that the nymphs do little to remedy the limited egg dispersal. If the eggs are laid in a cluster and the nymphs fail to disperse themselves, the young insects must compete with one another for plants to climb and leaves to eat. The habits of both mothers and nymphs seem ill suited to survival. How does the new generation escape predation,

ILLUSTRATION 3 With a forceful flick of
her abdomen, a stick insect throws an egg
several meters at an initial speed of about
3 meters (10 feet) per second.

find sufficient food, and survive? How do these insects efficiently
spread themselves to fresh plants and new areas?

The answers to these questions lie in the eggs. Stick insects offer
their offspring neither food nor care, but the eggs themselves provide for the next generation in a surprising way. The eggs are small
and enclosed in a hard shell rich in calcium, so that they resemble
plant seeds more than insect eggs and thus may deceive some
egg-eating predators. Some species' eggs are brown and shiny,
doubtless further enhancing their resemblance to seeds. Perhaps
even more important than its appearance is the appendage that
each egg bears. This appendage, or capitulum, is rich in lipids (fatty

compounds), and it can be detached without diminishing the egg's viability. It proves to be the key to dispersing the stick insects' eggs.

The secret of the capitulum is that it contains chemical compounds that attract ants. Foraging ants come upon stick eggs in the leaf litter. Attracted by the odor of the fatty capitula, they pick up the eggs and carry them home to their underground nests. There, they snip each capitulum from its egg with their strong mandibles and feed it to their larvae. Having no further interest in the egg itself, they either abandon it in the nest or perhaps discard it onto their garbage dump. In this way, the ants' food-gathering has effectively relocated the egg, and now it will hatch at some distance from where it was laid. By providing a nutritious bit of food, sedentary stick insects have paid ants to disperse their eggs. Not all ant species participate in this scheme, but enough do so to spread the eggs about.

Stick insects benefit from this arrangement with ants in other ways as well. The eggs hatch only after several months, and by passing this time underground they remain beyond the reach of destructive fires and aggressive predators. In particular, an ant nest is an ideal place for eggs to avoid the parasitic wasps that are perhaps their worst threat to survival. The eggs' seedlike appearance may deceive egg-eating birds, but parasitic wasps are not so easily misled. Probably hunting by odor, these tiny, brightly colored insects search for stick eggs as sites for laying their own eggs. On finding a suitable egg, the female wasp grasps it, gnaws a hole in its hard cover, and lays her own minute egg within. When the wasp larva hatches, it kills the developing stick insect, commandeering the egg and its store of food. The egg's chances of hatching are much better if it is hidden among the ants.

We know little more about the bargain between stick insects and ants. The capitulum's function as a transportation ticket is clear: Ants pick up and carry only eggs bearing an attached intact capitulum. The eggs of stick insects that do not depend on ants for dispersal typically have little or no capitulum. However, no one has yet identified the capitulum chemicals that attract ants and induce them to carry the eggs back to their nest. Although this story is incomplete for the present, we can pursue these signals to ants by examining a closely related situation. Stick insects are not the only creatures to buy transportation from ants, for many flowering plants make comparable arrangements for dispersing their seeds. This bargain between plants and ants has attracted attention for almost a century, and we can best continue our story by turning to it now.

Rooted to one spot and lacking mobility, plants need help in disseminating their progeny just as sedentary stick insects do. They handle this need in various ways. Some plant seeds have wings or hairs that allow them to sail abroad on the wind. A few bear small air sacs and so can float away on water. Fruit-eating animals incidentally consume seeds embedded in tempting fruits and later excrete them elsewhere. Other seeds equipped with hooks or spines become attached to passing creatures and so travel afar. Some seeds, perhaps less familiar than these other types, are designed to be dispersed by ants.

Like stick insect eggs, ant-dispersed seeds have detachable fatty appendages that attract and reward ants. The seed-appendages are called elaiosomes, but they function in much the same way as capitula. (Elaiosomes and capitula received their names independently, long before their analogous functions were recognized.) Ants gather these seeds because the elaiosomes' mixture of fat, protein, and even some vitamins provides excellent food for their larvae. The plants in return receive much the same benefits as stick insects do. Their seeds are dispersed to ant nests, where they are safe from fire and predators. The ants may also furnish a favorable site for growth by disposing of a seed in their garbage. Ant garbage dumps collect all sorts of organic waste, fecal matter, corpses, and remains of prey, and so can be full of nutrients, making them fertile places for a seed to sprout.

For plants, probably the most significant aspect of this arrangement is safeguarding their seeds from predation. In some cases, the ants' protection is not merely advantageous but definitely necessary for a plant species' long-term survival. Nowhere is this more apparent than in the botanically rich Cape region of southern and southwestern South Africa. This area is home to a large number of plants that are found nowhere else, and that are now threatened by an invasion of tiny Argentine ants (*Linepithema humile*). These foreign aggressors reached South Africa in freight and baggage at the beginning of the twentieth century, and by the early 1980s they had spread to the Cape region. Having made their way from South America around the world, they are notorious on several continents for their destructive behavior. In Southern California they are a serious agricultural pest and a household nuisance worse than fleas and roaches.

In South Africa, as in other areas they have colonized, Argentine ants promptly supplant the dominant local ants. In parts of the Cape region, indigenous ants were responsible for dispersing the seeds of more than 150 plants. As the local ants have retreated before the invading swarm, these dependent plants have begun to vanish. Unlike the ants they displace, Argentine ants make no bargains with plants. When an Argentine ant finds a seed with a fatty elaiosome, it simply eats the elaiosome and drops the seed to the ground. Seeds that were once removed to the safety of ant nests now remain in the open. Exposed to seed-eating predators, many of these seeds are now lost. The fate of a familiar Cape plant called red stump (*Mimetes cucullatus*) exemplifies what can happen. In areas where Argentine ants have established themselves, there are only two percent as many red stump seedlings as where local ants still prevail. As the Argentine ants advance, red stump is disappearing from the Cape, its seeds no longer secure from predators. Unless the invaders are stopped, the future of many Cape plants is uncertain. Biologists from South Africa to Southern California are looking for practical ways to control these destructive creatures.

Elsewhere, long-established arrangements between ants and plants continue unthreatened, and plants with ant-dispersed seeds flourish in many temperate and tropical areas. Overall, these arrangements involve more than three thousand species from more than eighty plant families. In North America, these plants include some trilliums, hepaticas, and violets, along with about thirty percent of other spring-flowering plants native to moist broadleaf forests. In Australia and southern Africa, on the other hand, elaiosome-bearing seeds typically belong to tough-leafed shrubs found in arid soils. In South America, the favored setting is again different. Here, ant-dispersed seeds come largely from plants growing in tropical rain forests. This diversity of plants and habitats implies that this plant-ant relationship has arisen independently several times during the 135-million-year history of flowering plants. At different times and places, flowering plants have again and again developed seeds with elaiosomes attractive to ants. These developments vary widely in their details, as you would expect for historically independent events. For example, elaiosomes in unrelated plants may originate from different parts of the seed, so even though all elaiosomes serve as food for ants, they consist of a variety of seed tissues.

A central feature of this plant-ant bargain are chemicals that induce ants to transport seeds back to their nest. Unlike the situation with stick insects, we know something about these signals from plant to ant. Scientists have examined only a few kinds of ant-dispersed seeds, but those examined have important features in common. In all cases, the chemicals that attract ants are lipids in the elaiosome. One effective attractant is a common substance called oleic acid. Ants pay particular attention to seeds that have relatively large elaiosomes rich in oleic acid, eagerly grasping and picking up these seeds. This response to oleic acid is not surprising, because ants are familiar with this compound as a signal within their own species. Oleic acid is a signal for many different ants (and also bees), and it carries the same message for all of them. Apparently, many plants have appropriated this popular ant signal for their own purposes. Let's look briefly at how ants use oleic acid.

Oleic acid is one of the "monounsaturated fatty acids" that nutritionists talk about. (Fatty acids are acidic compounds that take their name from their origin in the decomposition of natural fats.)

Oleic acid is a normal constituent of animal fat, including ant fat. When an ant dies and its body begins to decompose, its fat breaks down and releases odoriferous fatty acids. If the ant dies within its nest, the odor of oleic acid serves as a posthumous chemical signal to its surviving nestmates. On detecting oleic acid, an ant worker's response is to pick up the source (the dead ant) and carry it a short distance toward the nest entrance before setting it down. Eventually, after several workers have moved it, the carcass reaches the entrance, where it is finally ejected from the nest.

Plants have apparently adopted the dead-ant signal to induce ants to pick up seeds that contain the signal. Evolutionarily, this development should have involved a minimal cost in energy, since plants were already making and using oleic acid as a constituent of their fats. Here then, the critical step would have been not the synthesis of oleic acid, but the incorporation of oleic acid into the seed's elaiosome.

An ant picks up an oleic acid–containing seed just as it would a dead ant. However, instead of simply moving the seed a short distance and then dropping it, the ant takes it back to the ant nest. This second step implies a second signal, and several different compounds seem to fulfill this role in different seeds. In some seeds, the second signal appears to be another common fatty acid also present in elaiosomes, called linoleic acid. For some kinds of ants, linoleic acid serves as a feeding signal. These ants associate the odor of linoleic acid with food. Alternatively, in other plants the second signal seems to be compounds identified with ant brood (eggs and larvae). This brood odor may stimulate ants to carry seeds home, just as they would eggs or larvae they discovered outside the nest. In several other seeds, a derivative of oleic acid called 1,2-diolein (1,2-D) is a very effective second signal. Insect fat contains 1,2-D, and to an ant this odor may denote an insect that it should carry home as food. Additional second signals almost certainly exist, and perhaps also third and fourth signals not yet recognized. This lack of specificity is not unexpected, owing to the multiple, independent origins of these arrangements between seeds and ants.

Another chemically mediated transaction between plants and ants that is more complicated and much less common than seed-dispersal appears to be limited to a few species of tropical tree-dwelling ants and several small plants adapted to growing in trees rather than in the soil. The ants collect the plants' seeds, and the plants grow on, or really as part of, the ants' arboreal nests.

These nests are known as ant gardens, and one of the best places to find them is in the hot, wet forests of southeastern Peru. Here east of the Andes and on the western fringe of Amazonia, there may be an ant garden every 15 meters (50 feet) in trees along the forest trails. The gardens sometimes are essentially continuous, particularly in flooded or disturbed areas. Biologists have studied ant gardens since the early 1900s, but recent work in Peru has materially expanded understanding of them. Most Peruvian ant gardens are the work of two species of ants. One is a large aggressive creature called *Camponotus femoratus* and the other is a small one called *Crematogaster limata parabiotica*. These two species maintain an unusual and not very well understood relationship, sharing trails and sometimes food, and inhabiting interconnected nest chambers. They fashion their nests from earthen material, plant fibers, decaying leaves, and other organic debris, cementing these components together with their own glandular secretions to form nests that they attach to tree crevices and crotches high above the forest floor. To these nests they bring several kinds of seeds, scattering them in the earthlike material. The seeds germinate and mature into plants whose roots ultimately reach throughout the nest structure, binding this material together and lending it strength. The plants form a mass of thriving vegetation that almost completely conceals the nest itself and transforms the entire structure into an ant garden.

The plants do more for their ants than strengthen the nests. The most prevalent ant-garden plants bear moist, pulpy fruits and produce generous amounts of nectar from special glands. The ants can exploit these nourishing foods without leaving home. In return, the ants offer their plants a secure, well-tended site. They protect their plants' seeds from predation and most likely discourage marauding herbivores. From time to time, the ants add fresh nest material where the plants have sent out new roots, and they fertilize the garden by collecting vertebrate feces and incorporating them into the nest material. Plants flourish in this rich environment, and indeed the commonest ant-garden plants are rarely found growing elsewhere.

For this arrangement to succeed, the ants must choose appropriate plants. Not every plant species can thrive in an ant garden, nor would every plant offer the ants suitable food. Ants must select their plants with as much care as those human gardeners who pore over nursery catalogs all winter deciding on the seeds to buy for spring. Because some seeds the ants select for their gardens have elaiosomes, we might assume that elaiosomes are a particular attraction. However, "cafeteria experiments" indicate otherwise. In these experiments, Peruvian ants selected seeds for their nest from offerings that included many local species (hence the descriptive term *cafeteria*). They rejected some seeds with elaiosomes and chose some without, indicating that elaiosomes are not a major consideration in their choices.

From recent work, we do know that chemical signals from the seeds direct the ants' selections. Although the story is still incomplete, an extraordinary finding has been that a chemical compound called MMS (methyl 6-methylsalicylate) is present in the seeds the ants choose. MMS is found in the seeds of the ten or twelve unconnected species from seven plant families that flourish in Peruvian ant gardens. The ants seem to find this compound irresistible. Offered inert particles coated with MMS, they become excited, pick up these decoy seeds, and occasionally carry them about.

In addition to MMS, the natural seeds preferred by ants contain other volatile compounds. Ants respond to some of these, and some are present in three or four different seeds. This may mean that the seeds emit several chemical signals, or that the necessary signals have multiple components. However, MMS may also hold other meanings for the ants. There is a trace of MMS in male *Cam-*

ponotus femoratus mandibular glands, suggesting that this compound may be a component of the ants' own odor. If so, the ants conceivably mistake MMS-containing seeds for their brood and carry them home.

The seeds chosen for ant gardens also may contain antifungal compounds that inhibit the growth of fungi where the seeds are planted. The fact is that destructive fungi do not invade ant gardens, although they infest neighboring piles of natural debris. There are also chemicals repellent to other ants. The Peruvian forest shelters ants that do not keep gardens themselves, although they are close relatives of ant-garden ants. As these species forage for food, they sometimes come upon ant-garden seeds. Unlike their ant-garden relatives, these ants find the seeds repulsive and refuse to pick them up. The seeds carry positive chemical messages for ant-garden ants and negative messages for others.

Peruvian ant gardens depend on a remarkable interaction between plants and ants, in which chemical signals from plants control ants' selection of seeds for their gardens. The advantages to both participants are easy to see, but we may wonder how this arrangement arose. How did a dozen independent plants start communicating with ants? To simplify a complex question, we can consider a single important issue: How did all these plants acquire MMS? This compound that the ants find so enticing is not an everyday natural chemical. Unlike the oleic acid and 1,2-D found in the elaiosomes of ant-dispersed seeds, MMS is rarely found in plants and animals. In addition to being in the mandibular gland secretion of one of the ant-garden ants, MMS turns up as a pheromone bearing various messages in several other unrelated ants. Two beetles use it for chemical defense, and a single microorganism, *Mycobacterium fortuitum*, synthesizes it, possibly in the course of making a growth-promoting agent. No other organisms that have received chemical scrutiny—no animals or bacteria, no algae or fungi, and no other green plants—appear to synthesize or use MMS for any purpose. Of course, millions of living species have escaped examination so far, but in the familiar biological world, MMS is rare. Yet, a dozen ant-garden plants in one small area of Peru incorporate this ant-luring chemical into their seeds. How did they do it?

This puzzle has several possible answers. Most biologists would probably consider ant-garden plants as a spectacular example of what is called convergent evolution. That is, each plant that incorporates MMS in its seeds has independently responded to a particular environmental situation in a similar way. Life in ant gardens offered such advantages that each species of ant-garden plant separately established the synthesis of MMS, which was somehow attractive to the ants. The notion that the extra biochemical machinery needed to synthesize MMS arose independently in a dozen unaffiliated plants all living in the same area may seem highly improbable. That an organism should evolve chemical communication with another species is remarkable enough, but that it should happen repeatedly in the same way seems very unlikely.

We do know, however, that unlikely events are not unusual across the evolutionary history of life on earth. Both fishes and whales glide gracefully through the water with minimal effort, owing to their streamlined, fish-shaped bodies. Despite the superficial resemblance of fishes and whales, this fish shape arose quite independently in these two groups at times roughly 300 million years apart. Fish-shaped fishes have swum in the earth's waters for 350–400 million years, long before mammals of any sort existed. Archaic whales, on the other hand, developed from land mammals and took up an aquatic life only 60–70 million years ago, independently rediscovering the advantages of a fish shape for life in the sea.

Convergent evolution is not the only conceivable explanation for MMS, however. An alternative possibility is that MMS evolved separately in each species of plant in response to environmental pressures that may have differed in each case but had nothing to do with ant gardens. Some time thereafter, the ants discovered each of these plants and, in response to the attractive MMS present in each of them, they brought the plants' seeds to their nests. In this way, all MMS-bearing plants eventually became established in a single location. At present this explanation seems less satisfactory than the preceding one requiring convergent evolution, because it demands for each plant specific environmental conditions favoring synthesis of MMS. Yet, no one has identified such conditions for even one of the ant-garden plants, and we are aware of no other plants elsewhere that make MMS for any purpose.

A third possibility is that MMS is produced not by the plants themselves but rather by a microorganism that infects the plants growing in an ant garden, and the plants then incorporate microbe-synthesized MMS into their seeds. This possibility demands the least novelty and avoids a dramatic instance of convergent evolution, but it postulates a microorganism no one has yet discovered. All three possibilities for the origin of ant gardens have received serious consideration, but as yet no way has been found to decide among them or other possible explanations. Ant gardens remain imperfectly understood, but along with the arrangements that plants and stick insects make with ants, they stand as extraordinary examples of mutual interspecific accommodation mediated by chemical signals.

CHAPTER

Getting Pollinated

3

Just as immobility forces plants to make special arrangements for dispersing their seeds, it also complicates their reproduction. No less than animals, flowering plants depend on bringing sperm and egg together to propagate their species. Their inability to move about presents the problem of how to bring these reproductive cells together. A plant's sperm cells are carried in pollen grains that ripen on parts of a flower called anthers. Pollen must be transferred to another part of a flower, the stigma, before sperm and egg can come together. In some cases, a single plant or even a single flower might pollinate itself, that is, move pollen from an anther of one flower to the stigma of the same or a nearby flower. This is possible at times because individual plants of some species bear both male and female flowers or a single flower may afford both sexes. However, for most plant species where it is possible, self-pollination—or inbreeding—is not a realistic option, as it does not lead to fertilization or at best furnishes inferior progeny. To set good seed, most plant species avoid inbreeding and require pollen from a different individual.

As a result, plants have a quite different focus to their reproductive activities from that of most familiar animals. From butterflies to bowerbirds, animals pour great energy into sexual rituals that have one primary purpose: A male seeks to attract a receptive female and convince her of his superiority over competing males. To do this, perhaps he brings her a gift, wins a fight, or builds her a nest. By one means or another, he attempts to prove his excellence so the female will permit him to father her offspring. This face-to-face evaluation by the opposite sex has no place in the plant world. Here, successful reproduction is possible only if suitable pollen is somehow delivered from one plant to another, and so plants devote their energy to securing some third party's assistance in carrying out this task. Unfortunately, outside pollinators are not always easily obtained. For nearly half the species biologists have investigated, third-party assistance is the limiting factor in natural reproductive success. For these species, the critical issue is not the weather, predation by seed-eaters, or the fertility of the soil, but simply whether a sufficient number of pollinators appear at the right time. Without pollinators, there will be no next generation.

One solution to the pollination problem is to avoid dependence on other organisms and simply allow the winds to spread pollen about. This is the method found in cereal grains and other grasses, as well as in many trees. Because the wind can be erratic, it is most effective when plenty of pollen is available. An acre of corn (maize, *Zea mays*), for example, sheds about 130 kilograms (nearly 300 pounds) of pollen from its tassels over a brief period of one week. This is some 25 million pollen grains per cornstalk, or between twenty thousand and fifty thousand times more pollen than is needed to fertilize the acre completely. With so much pollen at hand, the wind is a thoroughly effective pollinator. Keeping in mind that each kernel on an ear of corn is a seed resulting from one successful event of fertilization by a wind-borne pollen grain, we can regard a field of ripening corn as visible evidence of how effective wind-pollination can be.

Wind-pollination works best when plants of the same species grow closely spaced, as in a field of grain or a stand of trees. Not many plants grow this way, however, and in fact fewer than 3 percent of flowering plants rely on the wind to disperse their pollen. Instead, most have long-standing arrangements with animals. In the main, these arrangements involve insects: More than 289,000 species of insects act as pollinators of flowering plants. Bees come to mind as the typical insects involved with flowers; however, long before bees were plentiful, beetles were well established and were probably the earliest insect pollinators. Flowers provide some beetles with snug nests, places to lay their eggs and raise their young, and convenient food. In the course of their daily activities, they inadvertently transfer pollen from flower to flower. Birds and bats are also significant pollinators, and even rats, mice, and monkeys serve a few plants.

Many undiscriminating plants recruit a variety of animals for this work, but a few rely on one particular species, as does a strange Australian orchid (*Rhizanthella gardneri*) that lives and blooms underground. As the small plant grows upward and matures, it forces open the earth above, and in this crack its deep red flowers bloom well beneath the surface. Pollination of these underground blossoms is the business of tiny humpbacked flies that belong to a family (Phoridae) known as scuttle flies, owing to their habit of scuttling along the ground. Few other local animals ever even encounter the plant, and any that stumble into its subterranean habitat fail to take pollen from its flowers. The flies, however, belong to a group (the genus *Megaselia*) that frequents rotting vegetation on and beneath the ground surface, and their habits bring them into the orchid's world. The little flies visit the flowers and come away carrying bright golden pollen.

Transactions between plants and pollinators go back millions of years and vary greatly from species to species, but they all have certain features in common. A plant's goal is to coax pollinators to call upon its flowers, so the animal can both pick up local pollen grains and leave pollen it has brought from afar. To attract attention, plants advertise their presence with color and fragrance. Taking advantage of pollinators' color vision, their flowers have inviting colors and special markings that guide visitors to the flower's important parts. Most pollinators also have a sense of smell, and flowers give forth fragrances to solicit a passerby's closer inspection. Attractive hues and fragrances depend on colored and odoriferous chemicals, and to promote themselves in this way plants must synthesize the appropriate chemical compounds. As we explore pollination, we shall be particularly concerned with the role of fragrant chemicals.

To encourage pollinators to call repeatedly, plants reward them for their visits. For many insects, pollen itself is reward enough. It is a nutritious food, highly proteinaceous and full of vitamins. Bees collect it avidly as a food for their youngest larvae. Of course, pollen collected by pollen eaters is no longer available for fertilizing flowers, so plants obligingly generate much more pollen than they need for reproduction. From a pollen collector's point of view, distributing grains of pollen from plant to plant is incidental to the important business of actively gathering pollen as food.

Pollen provides sufficiently good compensation that some plants, called pollen flowers, give away quantities of pollen and nothing else. Most plants, however, offer their visitors nectar as well. Nectar is primarily sugar-water, typically about a 40 percent solution, but it also contains amino acids and other nutrients. Cheap to make, it accounts for less than 5 percent of a plant's biochemical productivity. Although energetically inexpensive for plants, nectar is a significant source of food and energy for pollinators. A honey bee (*Apis mellifera*), for example, can gather nectar equivalent to nearly fifty milligrams of sugar (about a tenth of a teaspoonful) per hour. As an hour's sustained flight burns only about ten milligrams of sugar, a bee can accumulate a large amount of nectar for storage. For honey bees, storing nectar means concentrating it and mixing in some additives to make honey. In appreciating the prodigious amount of nectar bees collect, it is helpful to know that a single pound of honey represents the nectar from about seventeen thousand foraging trips and entails seven thousand bee-hours of labor!

ILLUSTRATION 4   Red ruffed lemurs
depend on traveler's tree nectar as an
important food source and drink it without
destroying the flowers.

Whereas a single drop of nectar may delight a bee, plants pollinated by vertebrates must be more generous if they are to satisfy
their larger visitors' appetites. The pollinator of the Madagascan
traveler's tree (*Ravenala madagascariensis*) is the ruffed lemur (*Vare-
cia variegata*), a small tree-dwelling primate native only to Madagascar. The tree has a striking fanlike shape and beautiful ten-foot-
long oblong leaves that have made it a popular garden showpiece
in Florida and northern Australia (where it is successfully pollinated by local bats). The tree's sturdy flowers contain large nec-
taries, each holding almost 7 milliliters (nearly a quarter of a fluid

ounce) of nectar and assuring frequent visits by the lemurs. Every time a lemur pushes its face into a flower to sip this rich, renewable liquid, its fur is dusted with pollen that it spreads from tree to tree.

Obviously, pollinators are indispensable to most flowering plants. What is more easily overlooked is that they are critical to our own survival as well. Agriculture feeds the world, and about two-thirds of the world's crops require visits by animal pollinators to set fruit and seed. Various kinds of bees pollinate 60 percent of these crop plants, honey bees being the most important single species in this regard. In the United States alone, their contribution to crop pollination is worth billions of dollars every year.

As we noted, one of the ways plants encourage pollinators' visits is through an attractive fragrance. The appeal of this scent frequently depends on deception. A flower's scent may imitate an insect's sex attractant, the odor of its food, or an odor associated with a site for laying eggs. Fly-pollinated plants often reek of the carrion that is the flies' primary food. The odor comes not from rotting flesh, however, but from the plants' own mix of chemicals. Orchids in the genus *Ophrys* that are pollinated by male bees and wasps carry a similar deception even further to maintain these insects' services. The orchids emit an excellent imitation of a female insect's sex pheromone, and at the same time display flowers visually and tactually resembling a receptive female. As the males eagerly explore this bogus female and attempt to copulate with her, the flower loads them with pollen for delivery to the next seductive orchid they discover.

Unlike the deceptive fragrances of orchids, which are quite advanced plants, the scent of such relatively primitive flowering plants as magnolias is not intended to deceive. Instead, such ancient flowers emit an attractive floral fragrance that is a mixture of simple volatile chemicals. These same chemicals also serve the plants as defense substances. They are either toxic or foul tasting, and so deter feeding by foraging herbivores. This connection between odorous and defensive chemicals suggests one way fragrance may have originated as an attractant for pollinators. Over time, insects may have learned to associate the smell of a primitive plant's defense compounds with the presence of flowers laden with pollen or nectar. Once this association had formed, foraging insects would follow the fragrance to locate these valuable rewards. Plants would then have gained a novel way to solicit pollinators.

This suggested association finds some support from a group of organisms even more ancient than the earliest flowering plants. When insect-pollination first appeared in flowering plants over 100 million years ago, a quite separate set of green plants had already been enjoying a similar arrangement with insects for the previous 100 million years. (The independent development of insect-pollination twice in 100 million years is another example of convergent evolution.) These organisms that preceded flowering plants are called cycads. Once numerous and varied, they dominated the plant world during the Triassic and Jurassic periods (225 million to 135 million years ago), when dinosaurs roamed the earth. Cycads are the palmlike plants in the background in dinosaur pictures and movies. They outlasted the dinosaurs, but today only a few species survive. These relics are confined to tropical and subtropical climates and are frequently mistaken for ferns or palms. The common California sago palm (*Cycas revoluta*), for example, is probably the best known cycad in North America. The interesting point about cycads regarding fragrance is that they exhibit the same connection between defense chemicals and olfactory attractants as primitive flowers do: The same chemicals are again used for both purposes. The fact that this connection exists in two separate groups of green plants makes its significance more likely.

Cycads' pollination by insects may be one of the earliest forms of insect-pollination. Like conifers, cycads bear their reproductive cells in cones rather than flowers. Individual cycad plants are either male or female, male plants having pollen cones and female plants, seed cones. Fertilization requires transfer of pollen grains from pollen cones to seed cones. The role of insects in this process has received attention in a cycad commonly known as the cardboard palm (*Zamia furfuracea*). This is a handsome horticultural plant with stiff blue-green leaves and cones 15 to 20 centimeters (6 to 8 inches) long. A single weevil (*Rhopalotria mollis*) is responsible for pollinating cardboard palms, whereas other weevils pollinate a large variety of cycads. Weevils constitute a family of about fifty thousand species of long-snouted beetles, which appeared early in the history of insects. As we noted before, they may also have been the earliest insect-pollinators of flowering plants. The cardboard palm's weevil spends its entire life feeding and reproducing on this single cycad species.

Weevils first invade a new male cardboard palm during a one- to two-day period when the pollen in a cone matures and the cone's odor is particularly strong. As its pollen ripens, the cone metabolizes large reserves of starch and lipid at such a high rate that its temperature rises two or three degrees above the surroundings. This higher temperature helps volatilize the cone's fragrance compounds and so enhance its odor and attractivity. The cardboard palm's fragrance compounds are primarily a simple chemical called 1,3-octadiene and a common compound called linalool (lin·'al·o·ol). The latter compound is a defense agent widely distributed in flowering plants and other cycads, and has the scent of fresh flowers.

For the weevils, the maturing pollen cone with its pleasant aroma is an inviting new host. Swarming over it, they bore into the interior and feed on its starchy tissue. Then they promptly mate, the females laying their eggs directly into the body of the cone. The larvae that hatch days later feed without stirring from the cone and soon pupate. When the new adult weevils emerge, they move quickly to find another maturing pollen cone and repeat their life cycle. The entire interval from copulation to emergence of a new generation is no more than seven to ten days.

Feeding and breeding in the cone leaves the weevils covered with ripened pollen. As they move in the search for a new male cone to assault, they also come upon female plants bearing seed cones ready for pollination. Although they produce little heat to augment its intensity, seed cones give off the same fragrance as pollen cones, so the weevils make a brief stopover to investigate. As they explore, pollen grains scatter from their bodies, incidentally pollinating the cycad.

The weevils do not linger to feed and breed in seed cones as they do in pollen cones, even though starch is plentiful here as well. This behavioral difference is not yet completely understood, but it seems that the cones of female plants contain an active poison called beta-(methylamino)alanine (BMAA). This substance is a recognized neurotoxin for mammals and is believed to affect insects as well. The male plants' pollen cones also contain BMAA, but in a bound form that renders it harmless, allowing the weevils to nest and feed with impunity. In contrast, seed cones contain free toxic BMAA that presumably repels weevils and causes them to depart after quick examination and inadvertent pollination.

Through the combination of fragrant attractant and localized toxin, the plants substantially control the weevils' behavior. Having different forms of BMAA in male and female cones, the cycads prevent the weevils from being destructive parasites (organisms that benefit from their host while causing injury) and turn them into beneficial pollinators. Responding to the cycads' chemicals, the insects pass their life feeding and breeding in pollen cones and visit seed cones effectively only to shed some of the pollen they carry. From the cycads' point of view, this is a sensible arrangement. Housed in the pollen cones, the weevils do no serious harm. They greedily consume male starch and tissue, but the damage is an affordable price for pollination. Yet, if the weevils stayed to nest in seed cones, they could destroy the plants' eggs rather than only fertilize them.

Insofar as the cardboard palm and its weevil provide a good model, early insect-pollination must have been a rather messy, essentially accidental process. As time passed, however, interactions between plant and animal became more refined. For a more elegant arrangement, we return to flowering plants and consider the pollination of an orchid by moths. Although chemicals are once again part of the story, several other, very different considerations emerge here.

The tallgrass prairie that covered central North America for thousands of years is now nearly eradicated. An incredible sea of grass, in places ten feet high, once extended over nearly 220,000 square miles of the midcontinent. This vast expanse is now reduced to about 2 percent of its original area. Only patches remain here and there: areas in eastern Kansas, some in the Dakotas, bits in Minnesota, Illinois, Manitoba, and adjacent states and provinces. The largest single undamaged portion is a 30-square-mile patch just north of Pawhuska in northeastern Oklahoma. Elsewhere, the prairie has yielded to wheat, corn, and people. The surviving highly fragmented portions still provide an extraordinarily complex ecosystem, furnishing habitat to a tremendous variety of living creatures and sheltering some of the rarest organisms found in the midcontinent.

Among the loveliest of these rare species is the western-prairie fringed orchid (*Platanthera praeclara*), a stately plant about 40 to 90 centimeters (1.5 to 3 feet) high that bears two dozen or more creamy white flowers. With their showy fringe and delicate fragrance, these orchids stand out prominently against the surrounding grassy prairie. They can be found sporadically across the upper Midwest and Manitoba, but they are threatened or endangered throughout their range. On a single summer day in 1996, for example, twenty orchid seekers taking part in the Nature Conservancy's Great Orchid Hunt tallied only 608 plants in the southeastern corner of North Dakota, where the largest remaining expanse of the orchid's habitat is found.

One significant threat to the fringed orchid's survival is loss of suitable habitat as tallgrass is plowed under and the prairie is transformed for farming or development. Another, less obvious, threat derives from the fringed orchid's means of pollination. Despite their lovely flowers and delicate scent, the plants seem to go unnoticed all through a summer's day. No hungry bees or flies

ILLUSTRATION 5 As of 1996 there
remained only 175 separate populations
of western-prairie fringed orchids.

sample their nectar or gather pollen. Only later, after sunset, do the
orchids' visitors show themselves. As darkness falls, the white blossoms remain clearly visible in the prairie moonlight and their fragrance intensifies, as is typical of night-pollinated flowers in general. This evening display is for the benefit of two inconspicuous
nocturnal moths (*Eumorpha achemon* and *Sphinx drupiferarum*), both
members of the largish family of hawk moths (Sphingidae), and the
fringed orchids' pollinators.

    At first it seems curious that the western-prairie fringed orchid
relies only on these two pollinators and that its lovely flowers lure
no other night-flying insects. Closer examination reveals why this
is so. Because the orchids' fragrance becomes stronger at night,
scent is doubtless an important factor in attracting the moths. If this
scent has components that are particularly significant to these
hawk moths, these components may also contribute to limiting the

number of species that come to the orchid. No one has analyzed the scent, so we lack direct information on this point. We do, however, have analyses of three other *Platanthera* fragrances. These three other orchids all emit a mix of several of the simple floral compounds common to so many different flowers. If we can generalize from these three related species, we may guess that the fringed orchid's scent is probably attractive to more insects than just the two hawk moths. Yet the plants remain dependent on these two animals and no others.

The explanation of this dependence lies in the structure of the orchids' flowers. Fringed orchids store their nectar in what is called a spur, the long tubular or conical structure at one end of this flower type. To sample a flower's nectar, an animal must be capable of somehow reaching to the bottom of the spur where the nectar is stored. A flower with a spur reserves its nectar for those insects with a feeding tube, or proboscis, long enough to reach its bottom. In general, this is not a severe limitation, as numerous insects (and some birds) explore flower spurs and enjoy their nectar. However, the fringed orchids' spurs are the longest of any North American orchid, reaching a record length of 6 centimeters (2.5 inches). Few insects can probe such depths, and so their visits to these plants would go unrewarded. The two hawk moth species, however, are equipped for the task. Each has a long proboscis designed to drain nectar from the bottom of fringed orchid spurs.

Other insects may be capable of reaching the orchids' hidden nectar, but only the two hawk moths are capable of transferring its pollen. This unique capability is directly related to the distance between the moths' eyes. The orchids' flowers are so constructed that they steer an oncoming insect into a position for investigating the spur. Each flower has a long, deeply fringed lip that serves as a runway to guide the approaching moth's flight. As the insect moves in closer, two vertical lobelike petals, one on each side of the lip, direct its head to the proper location for the proboscis to plunge into the nectar-containing spur. When a moth reaches this point, it hovers in place, inserting its proboscis into the spur and probing for nectar.

The orientation provided by the flower speeds a moth's search and also positions the moth's head relative to the flower itself; just as the moth moves in close to drink, its head touches against two pollen-bearing organs, one on each side. The organs are so located that their pollen is cleanly transferred onto the moth's eyes. When the moth completes its drink of nectar and departs, it is carrying the orchid's pollen affixed to its eyes. At the next orchid it calls on, pollen grains are brushed from the insect onto two receptive structures whose location again matches that of the moth's eyes. Only an insect having its eyes properly spaced can transport the fringed orchids' pollen from flower to flower.

Only these two species of hawk moth appear to meet the dual requirements of long proboscis and properly spaced eyes and, as far as is known, only these two species pollinate the western-prairie fringed orchid. Other insects either ignore the plant as unrewarding or else fail as pollinators. The closely related eastern-prairie fringed orchid (*Platanthera leucophaea*) makes similarly specific demands on its pollinators, but in this case the flower transfers its pollen to the moth's proboscis.

The western-prairie fringed orchid's reliance on so few pollinators has been a fact of life for millennia. Developing in concert, orchid and moth fashioned an interaction more elegant and demanding than that of the cardboard palm and its weevil. It is an interaction that offers the moths a source of nectar unavailable to others and favors delivery of suitable pollen to the orchids. However, this irreplaceable relationship now threatens the survival of these splendid plants. Hawk moths, along with other beneficial insects, are vulnerable to chemical pesticides. Continued spraying of crops means fewer moths. As the hawk moths' numbers dwindle, pollination of the orchids becomes problematic. In some areas, the number of moths is so low that now only a very small percentage of the plants are pollinated and set seed. Recognizing the plant's increasingly uncertain future, in 1989 the U.S. Fish and Wildlife Service (FWS) listed the western-prairie fringed orchid as a threatened species. FWS and other concerned agencies have prepared a final recovery plan for securing the future of rare tallgrass prairie species, but it seems unlikely to be operable soon. Only 10 percent of the needed three-million-dollar funding is in hand.

The connection between primitive defense chemicals and attractants with a floral scent indicates that flowers use their fragrances for more than soliciting pollinators. One curious example of a flower's fragrance serving as a defensive chemical and not an attractant involves a plant with two distinct scents: dwarf Jacob's ladder (*Polemonium viscosum*), a small plant in the phlox family (Polemoniaceae) with sticky fernlike leaves. Its flowers range from light blue to dark purple and smell either sweet or skunky. The two scents come from different parts of the flower. Only rarely does a single flower give rise to both scents, although a single plant has flowers of both scent types.

The sweet fragrance attracts *Polemonium*'s principal pollinator, a species of bumble bee (*Bombus kirbyellus*). The bees call less frequently upon the skunky-smelling flowers, apparently finding the skunky odor relatively uninteresting. Although they prefer the sweet flowers, they do not seem to favor one color over another or to pay attention to the amount of nectar a flower offers. One shade of blue or purple is as good as another, and blossoms with extra nectar get no special attention. As a group, then, sweet-smelling flowers receive the most attention and thus are pollinated most often.

The bees' preferences raise a number of questions. Why do they favor sweet fragrance over skunky? Do bees share our human dislike of skunky odors? What is the skunky odor for, if not to attract pollinators? If pollinators prefer sweet-smelling flowers, how do the skunky-smelling flowers successfully compete for pollen and why does the plant produce them at all? At present, our knowledge is sketchy, and the most we can say is that the skunky scent primarily provides a defense against herbivores. One of the problems that dwarf Jacob's ladder faces is ants that steal nectar from its flowers but contribute nothing to dispersing pollen. Naturally, plants will fare better by saving their nectar for useful pollinators rather than wasting it on thieves who offer nothing in return. The skunky odor deters ant attacks and helps keep the thieves away. As for the remaining questions, we simply cannot answer them yet. Sweet and skunky flowers have different functions, and apparently maintaining both is worthwhile for the plant.

If flowers find fragrance useful for defense, it should not be surprising that they find other applications for it as well. Without doubt, the strangest and best-studied instance of scent employed as more than an attractant comes from a group of orchids known as bee orchids. The relationship between these flowers and their pollinator bees offers a fitting final report about pollination. Bee orchids encompass more than 600 species scattered over several genera native to the New World tropics. These plants generate large quantities of scent chemicals, which they then present as rewards to their pollinators. The bees belong to a group known as euglossine bees, comprising about 175 brightly colored species in five genera. Unlike honey bees, they are not social insects living in complex communities with shared labor; a few euglossine bees are semisocial, but most species lead solitary lives. The most remarkable characteristic of these bees is that they have adapted themselves to collecting fragrance chemicals with great efficiency.

The group of orchids that the euglossine bees pollinate bear flowers that yield no nectar, and their pollen cannot be gathered as food. Instead, the flowers' fragrance glands generate large quantities of scent chemicals that gather in oily droplets on the plants' surface. Their pleasant fragrance attracts only the males bees of the one or two species that pollinate each orchid. Female euglossine bees are indifferent to these scents, making this an unusual instance of pollination exclusively by male bees.

The specificity of each orchid's scent for only a few kinds of bees is at first surprising. The entire group of orchids makes use of about sixty different fragrance chemicals, each orchid typically using about seven to ten components for its particular scent. Most of these compounds are among the widely distributed plant chemicals we have previously met in cycads and in other flowering plants. On being tested individually in the field, several of the chemicals effectively attract many kinds of orchid bees. Even with these general attractants as components, however, a particular fragrance appeals only to a very few species. This specificity comes about because some of a scent's components act to limit its appeal.

For example, three constituents of the orchid fragrances are the common plant compounds known as cineole, benzyl acetate, and alpha-pinene. Cineole is a component of about 60 percent of the orchid fragrances, and benzyl acetate is in about 25 percent. When tested singly with Panamanian euglossine bees, cineole attracted thirty-five of the fifty-seven species tried, whereas benzyl acetate attracted six. A mixture of the two compounds might be expected to attract all the species that responded either to cineole or to benzyl acetate, that is, at least thirty-five different kinds of bees. However, a mixture of cineole and benzyl acetate attracts only eight species. Clearly, attractivity is not simply additive: A mixture's signal is not merely the sum of its parts. A bee that finds both pure cineole and pure benzyl acetate attractive may be repelled by a mixture of the two. When alpha-pinene is added, the resulting three-component blend attracts only two species of bees. Both the compounds present and their relative amounts are important in determining attractivity. Laboratory mixtures that accurately reflect the composition of an orchid's scent appeal to the same few kinds of bees as the natural scent.

When a euglossine bee arrives at an orchid that solicits his attention, he sets to work. Instead of looking for nectar or pollen as other insects might, he focuses only on harvesting the fragrance. He begins his task by emitting a blend of lipids onto the droplets of oily scent secreted by the orchid. The bee's lipid blend comes from a gland in his head and acts as a solvent to facilitate gathering up the orchid's oily secretion. Using brushes on his front pair of feet, he scrubs the plant surface and mops up the mixture of solvent and chemicals. Just like a household mop, these scrub brushes can hold only so much liquid. After perhaps thirty seconds of rubbing, the bee has picked up as much as the brushes can carry. Moving back from the flower and hovering, he quickly transfers the accumulated perfume to storage containers in his legs. First, he wipes each front foot with a midjoint of the middle leg on the same side of his body. This joint supports a stiff comb of bristles that picks up the tiny droplets of material as it wipes through the foot brushes. Then, from the comb on the middle leg, the bee transfers the material to a storage container built into one section of his hind leg on the same side. This part of his leg is grossly enlarged and has on its surface a slitlike opening filled with fine feathery hairs. As the bee brings his middle leg back, the droplets hanging on the comb touch these hairs and are immediately soaked up and transported into the chamber within the leg. Cells lining the chamber's wall somehow separate the bee's own lipid solvent from the collected fragrance chemicals and return it to the gland in the bee's head. From there it can be recycled for more collecting. Removing the solvent also increases the space available for storing fragrance chemicals. This three-leg transfer takes much longer to describe than to accomplish. With the two sides of his body working in concert, the bee can shift everything from his front feet to his hind legs in only a few seconds.

After working for some time at these chores, a bee may begin to behave peculiarly. If he were a human, we might assume he was drunk. Some biologists suggest that he does become intoxicated handling the scent chemicals; others believe he is only fatigued. Whatever the cause, he tends to slip, fall down, and become sloppy in his movements. None of this causes him to quit, however, and he may spend as long as an hour gathering and storing the flower's perfume.

The orchids' single benefit from supplying this rich reward is, of course, pollination. To ensure that the orchid bees perform as desired, the plants have devised incredible strategies. There are trigger mechanisms that shoot a pollen packet at the bee with considerable force, knocking him down but ensuring that the packet sticks to his body. There are flowers constructed like the western-prairie fringed orchid, so that a bee picks up a compact mass of pollen on moving about his business. One orchid (*Coryanthes speciosa*) has a wonderfully complicated system of this sort based on the bee's falling into a small bucket of water while he is at work. Perhaps the bee falls because the flower surface is quite slippery, or perhaps intoxication trips him up. In any event, he falls backward into a bucket formed by the flower's lip and partially filled with a watery liquid from glands on each side. Unable to climb out the way he fell in, the bee soon discovers the trap's one exit. To escape, he must crawl along a narrow passage leading under both the flower's anther and its stigma. If he arrived empty-handed, a packet of pollen is stuck to his abdomen as he moves past the anther. If he brought pollen from elsewhere, it brushes off onto the stigma. As he follows the passage to safety, the bee pays for the fragrance he has taken and fulfills the orchid's need. Not every visiting bee tumbles into the bucket and must then crawl past anther and stigma, but enough do so that the scheme succeeds.

A few other kinds of flowers have adopted the orchids' trick of trading odoriferous chemicals for pollination. The bees also harvest chemicals from decaying, fungus-infected logs, rotting fruit, and animal feces. Chemicals from other flowers are similar to the ones from orchids, but decaying logs furnish substances not available elsewhere.

The bees must have some important role for these chemicals, for without them their lives are shorter. They possess remarkable structures for collecting and storing scents and devote enormous effort to the task. Why do they do this? What do they do with the chemicals they collect? Unfortunately, we know too little about euglossine bees to answer these questions definitively. It is easy to attract and trap male bees using synthetic fragrance chemicals such as cineole as bait, but females are considerably more difficult to find. As a result, several euglossine species are known only through male specimens, and much in these creatures' lives remains hidden from our view.

We do know that some male euglossine bees mark out territories, either individually or in groups, and then attract females to these areas for mating, presumably employing marking signals and sex attractants in these activities. Perhaps the bees mix the fragrance chemicals with compounds they synthesize to formulate one or both of these signals. Another possibility is that the chemicals serve as starting materials for synthesizing components of one or both of the signals. That is, the male bees might simplify the biosynthesis (biological synthesis) of signal components by starting from compounds they gather rather than from less complex building blocks available through metabolism. Experiments to clarify the fate of the carefully collected fragrance compounds are not difficult to imagine, but carrying them out on wild bees in a tropical forest will be demanding.

From the plants' point of view, of course, what the bees do with the chemicals does not matter. What is important is that the bees come to gather fragrances and in doing so fulfill the plants' needs. This unique reward system uses larger quantities of chemicals than those of other plants, but the consequence is just the same: As always, the plants' vital goal is getting pollinated.

Flies and the Misery They Bring

4

The insects we have been discussing all have good reputations, and all stand high in human regard. Stick insect fans even maintain a page on the Internet to proclaim warm enthusiasm for their retiring pets. Bees and ants pursue their busy lives in communities that have long symbolized industry and diligence. Apart from entomologists, however, many people find other insects disgusting rather than admirable. Among the more repulsive beasts, flies must surely be leading contenders for the most despised of all. Few people find flies engaging, and probably no one keeps them as pets. For biologists, this group of unpopular creatures includes not only such common insects as house flies and horse flies, but also others that are not called flies at all but have such familiar names as mosquitoes, gnats, and midges. All of these insects are what biologists sometimes refer to as true flies, which are the two-winged insects belonging to the order called Diptera. The name true flies is used to distinguish these creatures from various other insects that are popularly called flies of one sort or another but are not Diptera. We are familiar with butterflies, dragonflies, and house flies, for example, but of these three only the house fly is a true fly. Both butterflies and dragonflies have four wings and belong to other insect orders.

ILLUSTRATION 6 Although amber
fossils most commonly contain flies, animals
as large as lizards have been found.

Altogether, the Diptera include 125,000 species, all having two
wings, large eyes, and sucking, piercing, or sponging mouthparts.
In true flies, the second pair of wings common to other insects has
been reduced to two small knobs that assist in flying. As a group,
the Diptera are the best fliers among the insects and have spread
themselves around the world into both aquatic and terrestrial habitats from subpolar to tropical regions. The earliest records of their
existence come from fossils dating to the Triassic period (about 225
million years ago) and include individuals beautifully preserved in
amber. Some flies in amber appear essentially identical to species
now alive. These animals found their environmental niche early
and have exploited it successfully since long before birds or mammals appeared on the earth.

Flies, mosquitoes, and their kin probably deserve the bad reputation we give them. It is true that some serve as beneficial pollina-

tors of flowers, and that one large family of flies (Tachinidae) helps control undesirable insects by laying its eggs in pest larvae, which then serve as food for the fly hatchlings. Also, a small fruit fly (*Drosophila melanogaster*) is rightly famous as an experimental animal employed by generations of geneticists. Nevertheless, these real contributions to human welfare are overwhelmed by the annoyance and aggravation flies provoke, and much more so by the multitude of human and animal diseases they spread. Already in the classical world, long before any suspicion of their deadly role as disease carriers, flies and mosquitoes were infamous as tormentors of man and beast. According to the first-century encyclopedist Pliny the Elder, the Romans sacrificed to flies in the temple of Hercules Victor in the hope of winning some release from their aggravation. The Greeks are said to have sought relief through similar offerings.

The most familiar of these annoying creatures are house flies (*Musca domestica*), which are probably the world's most widely distributed insects. They are the animals most frequently associated with humans, for they have followed us to the ends of the earth. They are at home everywhere, breeding in manure, garbage, grass clippings, and almost any other organic material, and producing a new generation in less than two weeks during warm months. Each fly lives an average of a month and carries two to three million bacteria. Exploiting both filth and human food as they feed and lay their eggs, house flies physically transfer pathogenic bacteria from site to site and contribute to disseminating disease, particularly such intestinal disorders as diarrhea and dysentery.

In addition to plaguing humans directly, flies spread pathogens that infect a significant number of plants and animals. Veterinary medicine must deal with horse flies and black flies, as well as pests with such curious names as warble flies and bot flies. There are ubiquitous stable flies that look like everyday house flies but deliver a painful bite, and a multitude of less familiar species that prey on sheep, rodents, and rabbits. Flies that are crop pests typically feed on their plant hosts as larvae. Seedcorn maggots, for example, are the larvae of a small grayish fly (*Hylemya platura*) that infests seeds and roots of beans, peas, and other crops throughout temperate regions of the world. Hatching from tiny eggs laid in the earth, the maggots locate the germinating seeds that they feed on by the odors that the seeds produce.

Diseases borne by flies are serious, whether they affect humans, plants, or animals. The World Health Organization (WHO) estimates that diarrhea and kindred illnesses kill 3.5 million children annually. Flies contribute to this, but they also have much greater responsibility for another dozen or so diseases for which flies and mosquitoes serve as the essential vectors and that are caused by pathogens normally transmitted by these insects' bites. Malaria, yellow fever, sleeping sickness, and dengue are among the best known of these scourges. Although we see little of them in temperate climates, they afflict hundreds of millions of people, and literally billions of others are at risk, particularly those living in poverty in developing lands. Malaria kills more than two million persons a year, sleeping sickness and dengue, thousands each. Deaths owing to yellow fever are more difficult to estimate, as the infection is frequently misdiagnosed and significantly underreported. Caused by a virus carried by a mosquito (*Aedes aegypti*), yellow fever is a major public health threat in Africa and has been a serious medical problem in the New World since the slave trade brought it here in the sixteenth century.

None of these diseases is yet conquered, but one approach now being investigated is focused on chemical signals used by the flies and mosquitoes that carry them. If successful, such investigations should change the lives of millions and perhaps even influence the course of history, for epidemics of both malaria and yellow fever can have wide effects. Perhaps most notorious in the past century was the yellow fever epidemic that delayed construction of the Panama Canal. One hundred years earlier, yellow fever played a decisive role elsewhere in the Caribbean by hastening the French abandonment of Haiti. Following a decade of slave uprisings that had left the colony and its sugarcane plantations in turmoil, Napoleon sent twenty thousand troops to Haiti late in 1801 with orders to reassert French control. Under his brother-in-law, General Charles-Victor-Emmanuel LeClerc, the French army quickly subdued the rebellious slaves. LeClerc foresaw total victory as soon as promised reinforcements arrived, but before he could consolidate his gains, a fierce epidemic of yellow fever struck. Through repeated exposure over the years, the Haitians were resistant to the virus, but for the French, high fever and jaundice led all too often to delirium, coma, and death. Within a few months the disease had consumed a third of LeClerc's army and the general himself succumbed in the autumn of 1802. As fresh troops arrived, the epidemic raged on, ultimately killing tens of thousands of Frenchmen. Only about three thousand troops remained on the island when, a year later, Napoleon finally ordered the army's withdrawal. Its departure brought independence to Haiti and also marked the end of French aspirations in the New World. In a significant change of policy, France would no longer seek an American empire. Canada was already lost to the British, and soon after the reversal in Haiti, Napoleon accepted the young United States' offer of $15 million for the Louisiana territory, France's vast remaining North American holding. French imperial dreams in the Western Hemisphere had collapsed, and yellow fever was partly responsible.

The role of yellow fever in Haiti was not exceptional. Disease has always been a crucial if unpublicized factor in military operations, and fly-borne sickness has been a grave threat to armies for centuries. More than two thousand years ago, malaria contributed to a crushing defeat of Athens by Syracuse, and it has continued to influence warfare down to our own time. In the Second World War, the deployment of thousands of troops in the tropics resulted in devastating confrontation of American forces with this same plague. The peril first presented itself in the Solomon Islands, east of New Guinea, during the initial American offensive of the war, when the U.S. Marines invaded Guadalcanal on 7 August 1942. By the third week of August, malaria was rampant and cutting down the defenseless troops faster than the enemy. October saw

1,960 men hospitalized with the fever, and the number rose to 3,283 in November. Through 10 December 1942, American forces suffered 10,635 casualties in the Solomons; of these, 1,472 were from gunshot wounds and 5,749 from malaria. This grim threat soon received the attention it deserved, for such losses could not be sustained for long. The personnel eventually assigned to antimalarial activities in the South Pacific exceeded 4,400. Nonetheless, malaria caused five times as many casualties in this theater of war as enemy action. The U.S. Armed Forces suffered more than 460,000 cases of malaria before the war ended in 1945.

Public health strategies to combat malaria and other diseases transmitted by flies and mosquitoes vary as much as the diseases themselves, but one potential avenue of control now under investigation takes advantage of the insects' chemical signals. Just like bees and ants, flies depend on these signals for information about their surroundings. Many have chemical sensors built into their feet and can taste where they are walking. Black blow flies (*Phormia regina*) extend their proboscis and begin to feed when taste receptors on their feet are properly stimulated, typically by their walking on an animal carcass or other decaying organic matter. Many flies also initially locate food sources, such as victims to bite for a blood meal, through airborne chemicals detected in their antennae. In addition, flies employ airborne signals for finding mates and identifying appropriate sites to deposit eggs. The house fly, for example, releases a chemically simple sex attractant that has been prepared in the laboratory and is marketed for fly control under the name Muscalure.

Chemical signals associated with egg sites are probably widespread in insects, but we know much less about them than about sex attractants. These egg-site signals guide one of a female insect's critical decisions, her choice of appropriate places to lay her eggs. If larvae are to have a chance to survive, they must hatch into a suitable setting where proper nourishment is available. Not all larvae are as particular about their food as those of French truffle flies (*Suillia humilis* and related species), which feed selectively on the black truffles (*Tuber melanosporum*) native to the Périgord region of southwestern France. Growing underground in association primarily with oak and hazel roots, truffles are invisible from above. Gravid female flies locate these nut-sized fungi by their aroma, and then deposit their eggs in the soil nearby. When the larvae hatch, their first meal is close at hand. Truffles also appeal to many humans, who prize them as a gastronomic delicacy. (Black truffles sell in New York for twenty to thirty dollars an ounce.) Just as the flies do, local Frenchmen hunt truffles by odor, but they employ specially trained truffle pigs and truffle dogs to conduct the search. Black-truffle aroma consists of nine principal chemical components; trained pigs and dogs and untrained truffle flies respond to a suitable mixture of these compounds as avidly as they do to the complete natural aroma.

Another kind of egg-laying signal guides female southern house mosquitoes (*Culex quinquefasciatus*). These mosquitoes lay their eggs on water in large raftlike masses. About twenty-four hours after being laid, the eggs begin to emit a chemical compound that can serve as a beacon to other females seeking sites for their own eggs. Alone, this compound does not interest females, but if the water itself has the odor of an appropriate egg site, the combination of site and egg odors entices gravid females, encouraging them to add their eggs to those already present in the water. The combined signals from egg and site seem to denote a safe and suitable location for eggs. Water contaminated with rabbit feces contributes just the right site signal, implying that female mosquitoes look for polluted water to offer their offspring a proper start in life. Several simple chemical compounds ordinarily present in the feces of rabbits and other animals efficiently attract females when combined with the compound from eggs.

This egg signal may find some practical application. Southern house mosquitoes are common pests and can bear dangerous illnesses. In the southern United States, they carry two well-known parasitic diseases: St. Louis encephalitis, a human viral disease, as well as dog heart worm, a disease caused by a nematode (roundworm). In tropical regions, southern house mosquitoes are also host to a nematode known as *Wuchereria bancrofti*, which causes a parasitic disease of the lymphatic system called bancroftian filariasis, which afflicts more than one hundred million people living in the tropics around the world. The condition is frequently referred to as elephantiasis, owing to the grotesquely swollen and disfigured limbs that so often mark its victims. Filariasis is so debilitating that WHO regards it as one of the world's leading causes of permanent or long-term disability. In fighting bancroftian filariasis, the mosquito's egg signal could possibly be useful to public health officials. Its chemical structure is known and chemists could prepare it synthetically on a large scale. Together with site-odor chemicals, it could be sprayed over extended areas, inducing female mosquitoes to deposit their eggs in inappropriate places where larvae would not survive. It is not yet clear whether this would offer a practical and acceptable means of limiting mosquito populations and reducing transmission of filariasis.

Flies associated with other parasitic diseases also make use of chemical signals. Two of these disorders that are major public health concerns are leishmaniasis and sleeping sickness, both of which are caused by parasitic protozoa (relatively complicated one-celled microorganisms) belonging to a single family called Trypanosomatidae. About 350 million people are at risk for leishmaniasis and 55 million for sleeping sickness, virtually all in tropical or subtropical areas. The protozoa threatening these millions of people have life cycles that take them back and forth between flies and vertebrates. Although their parasites are closely related, the diseases themselves appear quite different clinically and illustrate different ways chemical signals can contribute to public health.

Leishmaniasis is rare in the United States, but it gained considerable publicity here after a small number of cases were diagnosed in American military personnel who served in the Persian Gulf during Operation Desert Storm. Each of these cases began with the bite of a tiny brownish sand fly infected with a protozoan parasite. There is no way of knowing how many other Americans sand flies bit and infected with parasites during Desert Storm, but no doubt others had infections that went unrecognized. Veterans' groups subsequently conjectured that undiagnosed leishmaniasis could account for some of the baffling complaints referred to as Gulf War syndrome.

Leishmaniasis is actually a complex of diseases caused by perhaps a dozen closely related organisms of the genus *Leishmania*, all of which are carried and transmitted by sand flies. Worldwide, about twelve million people are infected with these parasites, and about three million suffer from one of the clinical diseases. The various *Leishmania* species are virtually indistinguishable under the microscope and in early studies were sometimes confounded with the organisms that cause malaria. For years *Leishmania* were classifiable most readily by the form of leishmaniasis they induce, because their clinical manifestations are quite different. Three main forms occur in humans. The mildest is a cutaneous infection known as dermal leishmaniasis, or Oriental sore, usually localized at the site of the sand fly's bite. Kept clean, the sore will heal spontaneously in a year or less, and the patient is immune to reinfection. A more serious form of infection confined to Central and South America also begins with a lesion where the patient was bitten. It proceeds, however, to attack the mucous membranes of the nose and mouth. The resulting ulceration disfigures the lips, nose, and vocal chords, and secondary infections can lead to death. Finally, the most serious disease caused by *Leishmania* is visceral leishmaniasis, or kala azar (Hindi for black disease), which leads to progressive wasting with anemia and enlargement of the liver and spleen. Untreated, kala azar is fatal within two or three years.

At a time when little was understood about the leishmaniases, the complications of various indistinguishable *Leishmania* species and different clinical manifestations caused considerable confusion. The situation was further complicated both by the fact that most species of *Leishmania* infect at least one mammal other than man and by uncertainty in identifying the several species of sand flies that are *Leishmania* vectors. Taken together, these problems translated into slow progress in appreciating the nature of leishmaniasis and how it spreads. Classifying the flies remains an issue even today, but biochemical and molecular-biological techniques now permit more reliable identification of the parasites.

There is still much to learn, but there is a growing body of knowledge about *Leishmania* that infect humans. Like other parasites that move between unlike host species, these organisms have adapted themselves to more than one way of life. They spend part of their existence in a cold-blooded fly's alimentary system and part inside certain blood cells of warm-blooded humans, two environments that make quite different demands on an organism. While they live and multiply in a sand fly's gut, the parasites take the form of flat elongated cells, each 10 to 20 micrometers long (a micrometer is one one-thousandth of a millimeter; a small grain of salt is about 200 micrometers in diameter) and fitted with a whip-like flagellum at one end. They move about freely and reproduce by dividing in two, becoming so numerous that they block the fly's gut and push toward its mouth. Here they remain, poised to pass to a human host whenever the fly takes a blood meal.

As the fly feeds, some of the parasites pass into the human's bloodstream. Cells of the immune system immediately recognize them as unwanted intruders, and large white blood cells known as macrophages swoop in to engulf them. Normally, macrophages swallow up foreign cells and efficiently destroy them, but in this instance, after engulfing the parasites, the macrophages fail to kill them. Instead, a strange thing happens. After being swallowed by a macrophage, each parasite transforms itself from a long flat cell into a smaller spherical one and somehow blocks the macrophage's machinery for killing intruders. Safe inside the now disabled macrophage, the *Leishmania* cell begins to divide and redivide, soon filling the host cell with copies of itself. The macrophage swells to accommodate them, but, finally able to hold no more, it bursts and the spherical *Leishmania* cells erupt into the host's bloodstream. Here, fresh macrophages immediately take them up but are once again unable to destroy the parasites. The round of parasite division and macrophage destruction continues, with more and more parasitic cells bursting more and more macrophages. Depending on the species of *Leishmania*, the assault may extend into diverse parts of the body. The wholesale death of macrophages leads ultimately to the clinical symptoms of the various leishmaniases. If a second sand fly now bites the human host, it can ingest *Leishmania* cells with the host's blood. In the fly's gut, the spherical cells revert to the elongated form and *Leishmania*'s life cycle continues.

The parasites' extraordinary ability to survive within macrophages makes combating the infection difficult. It also raises a fascinating question about how *Leishmania* subverts a major bodily defense against foreign invasion, and this mystery has been the focus of numerous investigations. *Leishmania*'s secrets are now yielding to investigation, and the results are contributing to the development of vaccines against the leishmaniases. No suitable vaccine yet exists, but public health workers regard the possibility as a promising means of eventually controlling the disease. But although there is still no modern vaccine, folk use of vaccination against leishmaniasis has a long history. In Asia, "leishmanization," which is deliberate infection with the parasites to confer long-lasting immunity, is hundreds of years old and was used against cutaneous leishmaniasis as recently as the Iran-Iraq War (1980–1988). It is otherwise not widely employed and is useless against the clinically more critical kala azar. In the absence of a serviceable vaccine, treatment of leishmaniasis depends largely on classical chemotherapeutic agents containing the chemical element antimony. These drugs produce severe toxic side effects in some patients, and their toxicity, together with the difficulty of administering them and the long period of treatment required, has sharply limited their usefulness. They have not proved to be a practical weapon in the developing world where the leishmaniases are most threatening.

Improved chemotherapeutic agents are under development, as are other approaches to managing leishmaniasis, including efforts to learn more about sand flies and the chemical signals they employ. One way to eliminate the infection is to eliminate the flies that carry it, and this calls for detailed knowledge of how the flies live. Sand flies are small nocturnal creatures, 2 to 4 millimeters long (a millimeter is about one twenty-fifth of an inch, the width of about four large grains of salt, so a length of about a tenth of an inch). They are pale brown in color and covered with delicate hairs. They carry their diminutive wings erect above their bodies and, not being strong fliers, remain near the ground and move along in a series of wing-assisted hops. They comprise more than six hundred species in two genera, *Phlebotomus* in the Old World and *Lutzomyia* in the New. These species are widely distributed in the tropics and subtropics but also range as far north as southern Canada and the corresponding latitudes in Russia, some inhabiting animal burrows and wall cracks in relatively arid environments and others living in forest habitats. In the New World, no medically important species live farther north than southern Mexico, although in Mediterranean Europe sand flies carrying leishmaniasis reach from Spain to Turkey. Wherever they live, both males and females feed primarily on plant nectar and other sugar sources. Females of most species also take blood before laying their eggs. Using their very small, sharp mouthparts, they cut superficially into a host's skin, inject saliva containing compounds that increase blood flow, and then suck up the tiny pool of blood that accumulates in the wound. This blood meal provides nitrogen necessary for the eggs they will later lay in soil, leaf litter, or other organic material. (Nitrogen is an essential component of all proteins and nucleic acids, chemicals needed by all living organisms.)

Until recently little was known about chemical signals used by these flies. Much of our current knowledge in this area comes from research focused largely on a New World sand fly, *Lutzomyia longipalpis*, which is the only known vector for *Leishmania chagasi*, the parasite responsible for visceral leishmaniasis in Latin America. Over the past decade, this potentially fatal disease has emerged as a serious menace in northeastern Brazil. The fly is actually a species complex; that is, the name *Lutzomyia longipalpis* takes in closely related but not necessarily identical flies whose distinctions are not yet well defined. One of the most intriguing recent findings is a male sex pheromone that is chemically different in populations of flies taken from diverse geographic regions. Abdominal glands of male flies collected from three separate localities yielded different compounds, each of which is an active sex attractant for local flies in its region. Up to six different subpopulations of the species complex may exist, and it now appears that the chemically variable pheromone may assist in differentiating these populations more completely.

Male sand flies release this sex pheromone to attract females for mating. The males' attractant is more potent when mingled with odors from a host that can furnish a blood meal, so that a male sand fly is a more efficient lure for females when he is on an appropriate host. This host attractant in humans is some component of skin odor, but its chemical nature is still obscure. Experiments with human volunteers have revealed that individuals have widely differing levels of attraction for sand flies and that a single individual's attractiveness fluctuates over time. Male sand flies respond to these host odors just as females do, even though they do not feed on blood. In this way, flies can meet and mate on a host, and the mated female can proceed to take a blood meal at once in preparation for laying her eggs.

When it is time to lay these eggs, the female fly depends on chemical signals to guide her in choosing a site. Just as southern house mosquitoes do, female sand flies seek two kinds of signals: a pheromone from eggs already present in the site, and odors emanating from the site itself. The pheromone from the eggs originates with the female sand fly that laid them. As she lays her eggs, she coats their surface with this substance, which was recently identified as a common fatty acid (dodecanoic acid). Scientists have also identified compounds in rabbit and chicken feces that are effective markers for egg sites. These compounds are simple substances (hexanal and 2-methyl-2-butanol), but they differ from the fecal constituents that were found to attract mosquitoes. The signals from eggs and site strongly reinforce each other as attractants.

A nonchemical factor also influences a sand fly's selection of egg sites: She prefers narrow cracks and crevices rather than open surfaces. A gravid sand fly, then, is quite exacting about where she deposits her eggs. She avoids exposed areas and looks for a site that already contains eggs of her own species along with simple organic nutrients. Her concerns seem well placed, for these characteristics promise safety, protection from the elements, and food for newly hatched larvae.

It is too soon to say what role sand flies' chemical signals can play in the struggle to suppress leishmaniasis. It may be that signals marking egg sites could find practical application in stimulating egg-laying in inappropriate locations and thereby limiting the number of flies. Perhaps a synthetic combination of sex attractant and host odors could be sprayed over an area to confuse female sand flies and prevent them from finding both males for mating and hosts for a blood meal. Sex attractants of crop pests have been exploited in a similar manner for years. On the other hand, the growing information about chemical signals may be most significant as a contribution to our expanding knowledge of how sand flies live and interact with their environment. Investigators may turn up additional pheromones in *Lutzomyia longipalpis*, just as they have recently discovered a defensive chemical secreted by sand fly larvae. They are now also studying another fly in the same genus (*Lutzomyia pessoai*) that is implicated as a *Leishmania* vector in southern Brazil. Whatever application these pioneering studies find, in the long run they can only advance the struggle against leishmaniasis.

The second parasitic disease we want to consider is sleeping sickness, or African trypanosomiasis, as it is also known. Sleeping sickness results from an infection by protozoa called trypanosomes that are closely related to *Leishmania*, and, like leishmaniasis, sleeping sickness is spread by flies. On a more general level, however, the two diseases seem quite distinct. Leishmaniasis takes several forms, only one of which is fatal, but untreated sleeping sickness invariably leads to death. Leishmaniasis is a menace in much of the world but remains unfamiliar to a majority of Europeans and North Americans. Sleeping sickness, on the other hand, is an African illness, but has been notorious in the Western world since European exploration of central Africa began centuries ago. In addition to the human disease, a closely associated animal trypanosomiasis occurs and is known by its Zulu name nagana. Together, these two trypanosomiases have profoundly influenced the history of sub-Saharan Africa and severely hindered its development. Most concerned observers probably regard this backwardness as one more misfortune brought by flies, but not everyone sees the trypanosomiases so negatively: Some conservationists maintain that without these diseases, human encroachment would have long ago destroyed the rich collection of plants and animals that are native to equatorial Africa. Our present level of awareness should permit us to control trypanosomiasis while preserving these unique African forms of life.

The two trypanosomiases are endemic in a wide swath of central Africa, extending from coast to coast both north and south of the equator and covering an area larger than the United States. This territory is the extensive range of the tsetse flies (about thirty species in the genus *Glossina*) that carry trypanosomes associated with these diseases (*Trypanosoma brucei* and several related species and subspecies). Fifty-five million people in this region, mostly living in rural areas, together with uncounted domestic animals are at risk for trypanosomiasis. Occupying more than a third of Africa, the region covers thirty-six countries and accounts for half the continent's arable land. Historically, nagana has limited the usefulness of animals in agriculture and in moving trade goods throughout this vast territory. Cattle and sheep exposed to nagana may survive, but they do not thrive; horses, dogs, and camels simply die. More than 60 percent of the area would be suitable for either livestock production or mixed agriculture if the animal disease could be eliminated.

As for human trypanosomiasis, it is a major public health issue in twenty of the thirty-six affected countries. According to WHO, only 3 million of the 55 million people at risk are under surveillance, with the consequence that only a small fraction of the estimated 250,000–300,000 new cases are recognized and treated annually. About this same number of people die each year, and in badly hit villages 70–80 percent of the population may be ill. Without treatment, the fate of all these suffering people is inevitably an unpleasant death. The parasites induce a variety of cardiovascular, kidney, and endocrine disorders, and then, after two or three years, invade the patient's central nervous system. Two-thirds of the cases are identified only at this advanced stage, when the patients, no longer able to concentrate, become lackadaisical and indifferent to their surroundings. There is an initial period of aggressive behavior, but ultimately an insidious lethargy becomes overwhelming. Physical activity is out of the question. Eating and speaking, even opening their eyes, call for more exertion than victims can muster. They fall into a deep coma and soon die, usually of starvation.

Another dreadful aspect of sleeping sickness is that, in the absence of surveillance and control, it can quickly progress from an isolated case here and there to savage epidemic. Throughout its range, sleeping sickness regularly follows political unrest and social disruption, which are usually accompanied by neglect of public health and sanitation. During the Ugandan civil war, incidence of the disease in the country rose from fifty-two cases in 1972 to an average of one thousand new cases a year over the next eight years. In the Sudan, fifteen years of bloody warfare has brought a virtually uncontrolled epidemic of trypanosomiasis to the country's southern border. In this region adjacent to Congo (the former Zaire) and the Central African Republic, the incidence of sleeping sickness was less than 1 percent in 1989, following a decade of dedicated effort by Belgian physicians. With the arrival of Sudanese rebel forces in 1990, the Belgians fled. Continual warfare soon isolated the area, leaving the few local clinics poorly staffed and without medicine. Until Western charities returned in 1994, medical attention was simply unavailable. By that time an epidemic of sleeping sickness had taken hold, and today this epidemic continues to ravage the border region. In a single town, doctors have seen the number of cases rise from eighteen in 1995 to eighty-seven in 1996 to more than one hundred in the first half of 1997.

The trypanosomes responsible for this appalling affliction lead a complex life similar to that of *Leishmania*. As *Leishmania* do, they undergo changes of form on moving from fly to human and multiply by division in both hosts, although trypanosomes live free in the bloodstream of their human hosts rather than within macrophages. Whereas the two pathogens lead similar lives, the tsetse flies that carry trypanosomes are quite unlike the sand flies associated with leishmaniasis. Tsetse are more advanced flies, the size of a house fly or somewhat larger, with large strong wings that allow them to undertake much longer flights than sand flies can achieve. Even though they limit their daily activity to half an hour or less, tsetse are highly mobile. Moving perhaps as much as a kilometer (six-tenths of a mile) in a day, they are capable of reinfecting areas that have previously been freed of flies.

Both male and female tsetse live solely on vertebrate blood, and the various species that carry sleeping sickness typically feed not only on humans but also on both domestic and wild animals. Infected flies pass on trypanosomes whenever they take a blood meal, so that the parasites not only move between flies and humans, but also infect a number of other hosts. Infected domestic animals develop nagana, but wild animals may show no sign of illness. They serve instead as healthy animal reservoirs of trypanosomes, permitting tsetse flies to pick up the parasites at any time without necessarily feeding on infected humans or domestic animals. For this reason and also because available drug therapies have proved no more practical here than for leishmaniasis, control of trypanosomiasis has long emphasized eradication of tsetse flies.

In former times, eradication took the form of destroying the flies' savanna or woodland habitats, or slaughtering wild host animals. These indirect approaches were environmentally ruinous, and they gave way eventually to insecticides, sprayed either from the air or directly on the ground. In more recent years, attention has

shifted to trapping devices for both controlling tsetse and survey-
ing their populations. Visual traps simulating a host animal's color
and shape are effective for those flies that live along forest rivers,
but wide-ranging savanna species rely more on odor for locating
their next blood meal. As a result, the past twenty years has seen
intensive research on the chemical attractants in host odors and on
practical application of these chemicals in fighting tsetse.

This work began with the simple but important demonstration
in the 1970s that tsetse flies flock from afar to an ox tethered in a
roofed underground pit, showing that host odors are indeed effective attractants. Further experiments showed that increasing the
number of oxen used as bait increases the number of flies attracted
and that fattened animals draw flies more efficiently than starving
ones. Scientists then identified the most important attractants in ox
odor as two simple and very common chemicals, carbon dioxide
and acetone, along with a slightly more complicated compound
called octenol (1-octen-3-ol). For one kind of tsetse fly, mixtures of
these three components are nearly as attractive as natural ox odor,
but attractiveness differs from species to species. Cattle urine is also
a powerful attractant; much of its effect owes to two additional
compounds, 4-methylphenol and 3-propylphenol, both common
laboratory chemicals belonging to a class of compounds called
phenols.

These discoveries led to the simple traps that are now used to
monitor tsetse flies and to control them. A test operation in Zim-
babwe in the late 1980s gave dramatic evidence of how potent such
traps can be. Insecticide-impregnated screens, baited with acetone
and octenol, were distributed at a density of four screens per
square kilometer. These traps reduced tsetse populations by more
than 99.99 percent over an area of 600 square kilometers (230
square miles). The screens are even more effective when the urine-
derived phenols are added to the attractants employed. Their effec-
tiveness having been clearly demonstrated, traps and screens have
become important complements to insecticide-based measures for
controlling tsetse flies.

However, what works with one species of tsetse in one geographic area does not invariably succeed with different species elsewhere. Thus the search continues for feasible chemical signals. Several promising studies are now under way at the International Centre of Insect Physiology and Ecology (ICIPE) in Nairobi, Kenya. One of these concerns a tsetse species (*Glossina fuscipes fuscipes*) that carries human sleeping sickness but prefers to take its blood meals from monitor lizards (*Varanus niloticus*). Biologists at ICIPE are working to identify the attractant compounds in monitor odor in the hope of using them as bait for trapping these particular flies.

In another study, ICIPE scientists found that tsetse larvae emit simple oily compounds such as dodecane and pentadecane, which are also components of kerosene or light motor oil, and that these chemicals attract gravid female flies. The situation here is a little different from that of the egg attractants mentioned earlier, because tsetse flies reproduce in an unusual manner. Unlike most flies and other insects, tsetse flies do not lay eggs. Rather, fertilized eggs pass one at a time into the female's uterus, where they remain to develop and then hatch. Each newly hatched larva matures singly in the uterus for some days before the female deposits it on the ground. The larva then burrows beneath the surface and transforms itself into a quiescent pupa. Transformation of pupa into adult requires some weeks, the exact time depending on soil temperature. During this time, the larva's oily secretion attracts female flies ready to deposit other newborn larvae. Similar to egg attractants, this larval pheromone presumably encourages aggregation of larvae in sites safe and conducive to pupation and emergence as adult flies. The potential of the larval signal as a bait for trapping gravid females is now under investigation.

A third ICIPE study focuses on identifying tsetse repellents. Field biologists recognized long ago that tsetse flies bite some wild animals but not others. We now know that flies shun such animals as waterbucks, elands, and zebras, because the odors of these mammals contain fly repellents. This discovery precipitated a search for odor components that deter tsetse and led eventually to the identification of a specific repellent compound. Oddly, the compound identified is chemically related to the phenols in cattle urine that are tsetse attractants. Repellents could be useful in protecting humans and domestic animals from the bites of hungry flies, and chemicals of this sort are currently undergoing field testing. If both tsetse attractants and repellents become available for large-scale application, a push-pull strategy may be practical for clearing flies from a chosen region. A widely disseminated repellent could push flies from one area at the same time as an attractant pulled them into an adjacent one where baited traps could destroy them.

These ongoing investigations at ICIPE suggest that chemical signals have an enduring role in managing tsetse flies and controlling African trypanosomiasis. This is despite the disadvantage that the signals show some species specificity, so that each kind of tsetse must be considered and studied independently. However, chemical signals offer an offsetting practical advantage: Deployment of simple traps and screens built by local villagers in rural Africa is much easier and cheaper than spraying insecticides, which demands sophisticated equipment and rigid schedules to be effective.

Flies and mosquitoes are vectors for a number of other diseases that constitute significant menaces to public health, but we do not yet know what role chemical signals play in all of them. Where natural chemicals are important, much additional research will be essential to establish whether they offer realistic approaches to managing these afflictions. The results with leishmaniasis and African trypanosomiasis indicate that such research can both increase our understanding of how fly vectors live and also lead to practical measures for fighting the diseases they carry.

Eavesdropping as a Way of Life

5

Many creatures have found that getting on in the world is easier if they take advantage of chemicals belonging to others. Some carry this skill to a high art. In earlier chapters a few examples of their schemes have appeared: flowers whose fragrance imitates their pollinators' sex attractants; seeds that mimic ant odors to facilitate their own dispersion; and flies that follow the scent of truffles to locate egg sites suitable for their truffle-eating larvae. A multitude of other organisms rely on comparable schemes. Some, adopting the flowers' strategy, counterfeit other creatures' signals for their own purposes. Others usurp a genuine signal and turn it against its rightful owner. Another group quietly eavesdrops on neighbors' chemical messages to promote its own interests. Still others, like truffle flies, have devised ways to use their neighbors' defensive compounds or chemicals deployed for other purposes. Diverted use of special chemicals is widespread in the living world and essential to many lifestyles, and it deserves closer attention. The present chapter concerns eavesdropping on chemical signals, whereas the following one focuses on mimicking and stealing such signals.

A particularly large and varied class of eavesdroppers includes prey species that learn of impending danger from chemical signals disseminated by their predators. Prey responses to these warning signals are often entirely innate, requiring no previous knowledge of the predator. One striking example is the plainly terrorized reaction of pit vipers to common nonpoisonous king snakes. Pit vipers are a group of poisonous snakes (the subfamily Crotalinae) that includes rattlesnakes, copperheads, and cottonmouths (water moccasins)—three of the four general types of poisonous snakes found in North America. They take their name from a pair of heat-sensitive pits below their eyes that assist in locating prey.

Despite their own venom and reputation as killers, pit vipers such as timber rattlers (*Crotalus horridus*) and copperheads (*Agkistrodon contortrix*) normally respond with evident fear when confronted with a common king snake (*Lampropeltis getulus*). They hide their heads, thrash about, and display a stereotypical defensive behavior known as body bridging. In body bridging, a snake raises one or two vertical loops of its body and directs blows with them at an advancing enemy, a reaction utterly unlike the familiar coiling posture a viper assumes in other threatening situations. Even on its first encounter with a king snake, a viper displays this particular behavior. Further, there is no need for the two snakes' ranges to overlap so that they might routinely confront each other in nature. The response is an inborn defense, not a learned one.

The reason for the viper's fear is not hard to find. Pit vipers are on king snakes' list of desirable foods, for king snakes regularly feed on other snakes. Moreover, king snakes are immune to the vipers' venom and its enzymes, which kill by breaking down a victim's proteins. King snakes possess compounds that quickly bind to the destructive enzymes and inactivate them. For once, pit vipers are frightened prey. A hungry king snake simply grasps a viper by the head or neck and coils itself around the viper's body, suffocating its victim in a deadly embrace.

It is not surprising that a viper facing such a fate reacts defensively as soon as it picks up a king snake's scent. Snakes gather all sorts of chemical information by the frequent flicking of their tongues. Each flick picks up chemicals from the surroundings and transfers them to the mouth for detection. A viper's probing tongue quickly detects the complex scent of a nearby king snake, providing a first sign of danger. Among other components, this scent contains oily and waxy compounds from the king snake's skin, and it is these skin compounds that trigger body bridging and other defensive reactions. An extract of king snake skin frightens pit vipers as much as does the snake itself.

This response is an inherited trait, much as an infant's suckling is. Because such mechanisms are genetically based, they can endure long after ceasing to serve any practical purpose. Given the right stimulus, such behavior follows instinctively even after generations of irrelevance. A nice demonstration of a response persisting over a century of disuse comes from the behavior of Spanish wall lizards. There are two geographically separated subspecies of these reptiles, one (*Podarcis hispanica atrata*) native to the Columbretes Islands off the Castellón coast of eastern Spain, and the other (*Podarcis hispanica hispanica*) found on the nearby mainland. The mainland subspecies suffers predation by a local population of poisonous lizard-eating snakes known as snub-nosed vipers (*Vipera latastei*). These snakes used to be plentiful on the main island of the Columbretes but were completely eradicated during construction of a lighthouse there in the nineteenth century. Although several expeditions have searched for them, no one has seen a single snub-nosed viper on the island since 1886. For more than a century the island lizards have been essentially free of predation by vipers and other carnivores.

Nonetheless, the scent of snub-nosed vipers continues to alarm them just as it does their mainland relatives. Both lizard populations react with characteristic signs of stress, jumping abruptly here and there, alternately shaking their forelegs up and down, and lashing their tails from side to side. The century-long respite from predation has not dulled the island lizards' inherited fear of snub-nosed vipers.

Like pit vipers and wall lizards, various mammals are also sensitive to a predator's scent. Oftentimes, their best defense is retreat, and so frightened deer and squirrels withdraw quickly when they smell a predator. One odor frightening to prey mammals comes from the urine predators deposit to mark territories. Hunters apparently noted the effect of urine centuries ago and ingeniously took advantage of it. They learned to control the movements of game animals by sprinkling coyote or wolf urine in the forest, in this way maneuvering deer and other quarry into favorable hunting sites. Now, hundreds of years later, there is a modern counterpart to this hunting stratagem, for the aversion of prey animals to predator scent now sustains a flourishing trade in urine. One company marketing on the Internet at the address ⟨predatorpee.com⟩ offers gardeners and landowners "100% predator urine" to repel unwanted raccoons, deer, and "other critters."

It is not only vertebrates that benefit from eavesdropping. Much simpler organisms also exploit chemical signals belonging to others. A microbe's life is no less perilous than a mammal's, so it is not surprising that even single-celled creatures take countermeasures on sensing specific predator chemicals. One good place to observe this behavior is in the small organisms whose lives unfold in inconspicuous pools of water. Little pools contain a host of inhabitants possessing an efficient chemical sense, including an assortment of inconspicuous protozoa known as ciliates. These one-celled creatures, many measuring considerably less than one-tenth of a millimeter (one–two hundred fiftieth of an inch) in length, take their name from the mobile hairlike structures, or cilia, on their surfaces. Many reside in pools on the ground or in tree holes, where they feed on bacteria and other nutrients.

Often ciliates are the prey of insect larvae and other relatively large organisms that share their watery habitat. In western North America, tree-hole pools often harbor ciliates known as *Lambornella clarki*, as well as insect larvae that prey upon them. The larvae are the young of tree-hole mosquitoes (*Aedes sierrensis*), a pest species with the habit of laying its eggs inside holes in trees. The eggs hatch when rainwater fills the holes, and the nascent larvae then feed on *Lambornella* ciliates they find thriving in the water.

The larval feast may be short-lived, for the ciliates soon mount an amazing defense based on their capacity to alter their form and adopt a totally different way of life when mosquito larvae appear in their little world. The ciliates flourish as free-living cells when no predators threaten, but their lives change dramatically when they respond to a certain chemical that the larvae release. What this chemical is and why the larvae release it in the first place are still open questions, but its secondary message to the eavesdropping ciliates spells catastrophe for the larvae. On detecting the larval substance in the water, the ciliates undergo cell division, change their shape, and transform themselves from free-living cells into parasites. In their new form, they are able to invade and eradicate their attackers. The mosquito larvae's vulnerable prey has suddenly turned into a virulent parasite. After entering the larvae, the parasitic ciliates multiply until they finally kill their hosts. The dying larvae rupture and release numerous ciliates back into the water. These cells are the free-living form, but they can change themselves into parasites and then infest more larvae. By altering their form, the ciliates escape predation; they even succeed frequently in exterminating all the mosquito larvae in their pool.

The predators that feed on ciliates include not only insect larvae but also other kinds of ciliates. Many of the prey species detect their predators chemically, just as *Lambornella* detect mosquito larvae. One pair of ciliates plays out such a prey-predator relationship in the vernal pools prevalent in upper New England. These pools first appear in March as ice and snow start to melt, and as a rule they have vanished by July. Just after the spring thaw, active life gets under way as ciliates (a species of *Sterkiella*) emerge in newly formed pools. They will later become the target of predaceous ciliates (*Lembadion magnum*) that have not yet appeared. In the predators' absence, meanwhile, *Sterkiella* exist as flattish cells with a fringe of cilia and a smooth upper surface. When the predators finally emerge and start hunting them, *Sterkiella*'s structure changes markedly.

Here again, the predators themselves trigger this structural change by releasing a small protein that presumably binds to the prey ciliates, marking them for the predators' easy recognition. In a turnabout, however, the prey themselves have learned to recognize the predators' protein. They respond to its presence by developing keel-like protective projections on their body. Long, narrow bulges appear on each cell's smooth upper surface to fortify it against attack. Other sorts of prey ciliates protect themselves similarly with spines or lateral wings, usually generating these defenses without the necessity of cell division. *Sterkiella*'s new armor effectively deters attack: In comparative tests on cells in both the defended and undefended forms, predator ciliates consumed 73 percent of the smooth-form prey they captured but none of those armed with fully formed projections. Forty-eight hours' exposure to the predators' protein converts smooth ciliates to the completely armored form.

Why should these ciliates maintain two forms? Rather than switch forms when predators show themselves, why do they not retain their defenses all the time? After all, this is what many familiar creatures do. Snakes, spiders, poison ivy, and countless other organisms always remain armed. The answer apparently is that permanent adoption of the ciliates' defended state would come at too high a price. *Sterkiella* become biologically less fit on moving into their protected form. They grow sluggishly and their cell cycle is slowed down, with the result that in the defended state their numbers increase more slowly. If predation were a constant menace, this might be acceptable. Then perhaps only the protected form could assure continued existence, and reduced fitness would be the price of survival. But predator ciliates present only an intermittent danger, so there is no constant need for the less fit form. The overall advantage for ciliates seems to lie with shifting between a form that is biologically more fit and one that is better protected.

For many predators and their prey, the roles in eavesdropping are reversed. Prowling predators frequently take advantage of their prey's chemical signals as they search for food. These signals can be a vulnerable creature's downfall, as they unintentionally guide predators and parasites directly to it. One animal that betrays itself in this fashion is the five-spined engraver beetle (*Ips paraconfusus*), which is a common bark beetle and an economically significant pest in the western United States. A bark beetle flying through the woods searches for a tree suitable for colonizing. On discovering one, it lands and releases an aggregation pheromone to recruit more bark beetles to the site. Beetles receiving the message assemble to bore holes into the tree and construct galleries, where they lay their eggs. What bark beetles cannot appreciate is that their aggregation pheromone also attracts some of their worst enemies, including predatory beetles known as blackbellied clerids (*Enoclerus lecontei*). To these handsome gray and black predators, the pheromone signifies that bark beetles are congregating. Recognizing an opportunity to gorge themselves, they soon join the gathering throng, along with other forest residents with similar tastes. The predators feed, mate, and lay their own eggs in the bark beetles' galleries. In many cases, their larvae continue the assault by feeding on plentiful bark beetle eggs and larvae.

Instead of an aggregation signal, predators may recognize their prey's alarm pheromone. On receiving an electric shock, earthworms (*Lumbricus terrestris*) secrete a mixture that sends other earthworms wriggling away from the danger. When garter snakes (various *Thamnophis* species) perceive one of these worm chemicals, they flick their tongues rapidly and then strike. To the snakes, the earthworms' alarm pheromone announces that food is at hand. Similarly, wounded fathead minnows (*Pimaphales promelas*) discharge a distress signal from their broken skin. As the alarm spreads through the water, other minnows flee to safety, and eavesdropping predators converge on the wounded fish. Diving beetles (*Colymbetes sculptilis*) hurry to the scene, accompanied by young northern pike (*Esox lucius*) so voracious they consume one another as well as minnows. Interestingly, the minnows have found a way to profit from this abuse of their pheromone, for a wounded fish's signal is not a totally selfless warning. In their frenzy to reach an injured minnow, the newly arrived assailants begin to fight among themselves, striking one another and churning up the water. Often they create so much confusion that the injured minnow manages to escape.

A predator or a parasite may also exploit chemical weapons of its prey or host. Turnips (*Brassica rapa*) owe their distinctive flavor to sulfur-containing defense compounds, as do broccoli, horseradish, and other members of the cabbage family (Brassicaceae). One of these sulfur compounds called sinigrin makes turnips distinctly unpalatable to most herbivores. However, there are tiny greenish aphids (*Lipaphis erysimi*) that are turnip parasites, and sinigrin affects these insects quite differently. They like its taste. Far from being a deterrent for them, it is a powerful stimulant that promotes feeding. On finding a sinigrin-containing plant, turnip aphids cluster on the underside of its leaves and suck its sap. Not only does sinigrin mark their preferred food, but they also appropriate it as a chemical building block. Turnip aphids biochemically convert ingested sinigrin into an important component of their own alarm pheromone.

Many hunters react to their prey's overall scent or some of its components, perhaps the smell of fur or some less complex odor. One of the world's most injurious insects, the African malaria mosquito (*Anopheles gambiae*), prefers humans to other sources of a blood meal. Oddly, whenever possible the mosquitoes bite people on their feet. This predilection reflects their strong attraction to the mixture of fatty acids that we associate with smelly feet. Humans may find the odor offensive, but these mosquitoes know it as a fragrant guide to blood. The same fatty acids also draw them to another odor that offends some people, the smell of Limburger cheese.

Another group that locates hosts by means of a complex scent are flower mites. (Mites are small eight-legged parasites whose closest relatives are ticks, spiders, and scorpions. Although mites and ticks may resemble insects at first glance, the two groups are only distantly related.) One species of these mites is a diminutive creature perhaps half a millimeter (one-fiftieth of an inch) long called *Proctolaelaps kirmsii*. These tiny parasites spend their lives in the scarlet tubular flowers of firebush (*Hamelia patens*), an evergreen shrub common in tropical America. When a firebush flower begins to fade, the mites abandon it and wait about on the bush for another one to open. As a new flower unfolds just after midnight, they rush in to feast on pollen, enjoy nectar, and engage in an orgy of mating and egg laying. The mites depend on firebush for their existence, but they offer their host nothing in return and actually injure it. Their scurrying about does not pollinate the flower, because in firebush self-pollination is unproductive. Moreover, three or four dozen mites devour enough pollen to jeopardize the plant's ability to reproduce, and they consume enough nectar to diminish substantially the number of visits by its hummingbird pollinators. Hummingbirds call to drain the deep blossoms of their nectar. As they drink, they receive a dusting of pollen they carry away and transfer to other flowers.

These particular flower mites live and feed only on firebush. They move about from flower to flower but are too small to crawl from bush to bush. Most pass their entire two-week life on a single plant, but a few adventurous mites pursue their fortunes elsewhere. Perhaps they wish to search for more plentiful pollen and nectar, or perhaps it is the need for more mates that compels their departure, since both males and females must mate repeatedly to reach their reproductive potential. Whatever the impetus, some mites have the urge to travel and rely on hummingbirds to provide their transportation. In feeding, a hovering hummingbird inserts its bill deep inside a flower and laps nectar with its long tongue. Its visit lasts no more than five seconds, and during this time a mite wishing to leave the flower must race up the bird's bill and into its nasal cavity. There the mite can ride in safety, sharing the space with other roving mites as the bird makes its rounds from flower to flower. Most rides are brief, but at least one mite species (*Rhinoseius*

ILLUSTRATION 7 Together, flower mites
and hummingbirds consume 85 percent of a
firebush flower's nectar.

epoecus) makes lengthy journeys. They travel with migrating hummingbirds, spending the summer on the California coast and wintering in west-central Mexico.

A firebush mite has a serious problem in bringing its hummingbird ride to a successful conclusion. These little creatures remain faithful to firebush as their only host, although in the laboratory they can live on pollen and nectar from other plants if forced to do so. They specialize in a single kind of plant most likely to simplify locating mates and can reliably find their own kind only in firebush flowers. Should a mite land on the wrong type of flower, it will have no opportunity to reproduce. It is important then to get off the hummingbird at the right stop. As in boarding, the mite has only the few seconds of the bird's visit to descend and safely enter a flower. First it must be certain that the hummingbird is stopping at a firebush blossom. If so, it must then dash out of the bird's nostril, down its bill, and into the flower before the bird flies away.

How does the mite decide where to get off? Good eyesight guides hummingbirds in their search for nectar-bearing flowers, but flower mites are blind and their world experience comes through tactile and chemical signals. It is through its acute chemical sense that a mite decides whether to alight. Odorless to humans and hummingbirds, to a mite firebush flowers have a distinctive scent that sends it scrambling from its hummingbird at the appropriate time. The traveling mite samples flower scents and other chemicals each time the bird inhales. Every breath sends a gust of air through its nasal cavity and over the mite, furnishing a fresh bit of the local chemical milieu about one hundred times a minute. The firebush aroma must be quite characteristic, because only one mite in two hundred rushes into the wrong plant. Even with this accuracy, the mite still has a problem. In abandoning the bird, it must move fast or risk missing the flower by jumping off too late. As it races down the bill, the mite moves about 6 millimeters per second or just under 22 meters (about 70 feet) per hour. This may seem slow, but for an animal only half a millimeter long, it translates into 12 body-lengths per second, putting firebush mites in the same league as cheetahs running at top speed.

A less familiar group of eavesdroppers are insects that exploit other creatures as hosts for their eggs. They lay one or more eggs in some developmental form of a host, and this site then becomes both home and food supply for their larvae. The details are exceedingly variable, but in general the larvae ultimately kill the host or at least render it incapable of reproducing. Insects following this lifestyle are known as parasitoids. Like parasites, they live at the expense of other species, although there is a distinction between the two. Parasites live and reproduce with their host without immediately killing it, whereas a single generation of parasitoids kills its host and moves on to adopt a free-living existence elsewhere. Because their hosts include many economically significant pests, some parasitoids are sold in bulk for biological pest control in agriculture and forestry.

The parasitoid way of life has been enormously successful. There are many parasitoid flies, and parasitoid wasps are said to number more than one hundred thousand species. (Flies and wasps, along with butterflies, moths, and many other kinds of insects, develop by what is called complete metamorphosis; their eggs hatch to larvae that grow and develop, then encase themselves in cocoons, become inactive pupae, and finally emerge as adult moths.) There are even hyperparasitoids that lay eggs in other parasitoids, generating complex interactions among three diverse species. These creatures often escape casual notice owing to their small size. Many are no more than a few millimeters long, and some are much less. These are successful insects, but their lifestyle presents an inherent problem. An adult female must lay her eggs in the proper place if they are to be viable, and for many parasitoids this is only a single host species or a few closely related ones. As an adult, however, the female may never have seen her host, because she had left it before reaching that stage of development. She obviously requires a reliable means of recognizing her host. Parasitoids that lay only a single egg in each individual host face this problem again and again.

In reliably locating hosts, many females depend on odor. It may be the host's own scent or that of some associated feature that guides her. Tiny parasitic wasps called *Biosteres longicaudatus* lay their eggs in the larvae of Caribbean fruit flies (*Anastrepha suspensa*), which they find by following the strong smell of rotting fruit where the larvae mature. Three simple chemical compounds from the fruit (acetaldehyde, ethanol, and acetic acid) are particularly enticing to the wasps. These chemical markers, themselves products of microbial fermentation, are formed as bacteria and fungi feed on the fruit and decompose it. In this case, then, the feeding of one group of organisms (microbes) on another (fruit) yields a chemical signal that leads a third group (wasps) to the location of a fourth (fly larvae). Only with this elaborate assistance are the wasps able to reproduce. Such complex chemical interdependence of neighboring creatures is doubtless more prevalent than we realize.

A second parasitoid wasp called *Bracon mellitor* is an important weapon against a major agricultural pest. This wasp lays its eggs on the larvae of boll weevils (*Anthonomus grandis*), which are key pests on cotton. These wasps are the boll weevils' most important enemy in the southeastern United States and have occupied agricultural scientists for many years. A female wasp has the demanding task of locating larvae 10–12 millimeters long (half an inch or less) that are buried in cotton buds and bolls. A wasp's first indication of a hidden larva comes from chemicals released by a weevil-infested cotton plant. Like other plants, cotton responds to herbivores' attacks by launching a retaliatory chemical defense. Odorous components associated with this defense attract a female wasp seeking egg sites. Once she reaches a suitable plant, she pinpoints a larva through the odor of its distinctive yellow-orange excrement. The odoriferous chemicals that guide her are acid-derived compounds known as esters, which in this case are formed by linking several different fatty acids to cholesterol. On reaching the odor's source, the wasp stops to examine the area. If she is satisfied that a boll weevil larva lies buried below the surface in a bud or boll, she drills into the plant and then on into the larva. For this task she has an organ that combines her egg-laying apparatus with a separate sting. When her drilling comes upon a larva, she quickly immobilizes it with a shot of venom injected through her sting. Then she retracts the sting and deposits a single egg about 2 millimeters long. The wasp now withdraws completely, leaving her egg attached to the paralyzed larva, where it will grow and develop. She completes the entire operation in only a few minutes.

Some parasitic wasps are much less particular about hosts for their eggs. The larvae of *Trichogramma evanescens* develop successfully within most moth and butterfly eggs. Given such a wide choice of egg sites, it would be advantageous for some extensively shared odor to guide a gravid wasp. In fact, she is drawn to several waxy compounds, such as one that chemists know as tricosane, that are generally present on the tiny scales that cover moth and butterfly wings. Where there are adult hosts, host eggs are likely to be present within which the wasp can deposit her own eggs. Once she comes upon an adult host, the waxy attractants seem to stimulate her search for host eggs.

There is also a parasitoid fly that shows promise for controlling infestations of red imported fire ants (still nearly always called *Solenopsis invicta*, although since 1995 the correct name has been *Solenopsis wagneri*) in the southern United States. These small intruders from South America have become some of the most ferocious pests in the Southeast over the years since their arrival. They were probably stowaways in cargo landed at Mobile, Alabama, in the late 1930s, and from this point of entry they have spread both east and west. In sixty years, red imported fire ants have established themselves across the South from the Atlantic coast through eastern Texas, ranging as far north as Tennessee and southern Virginia. A fire ant's sting is painful to humans and livestock alike, causing an intense burning sensation followed by formation of a pustule that may become infected. For the 1 to 2 percent of particularly allergic people, a sting can be fatal.

The United States currently has an estimated 10 billion colonies of fire ants, many comprising more than fifty thousand workers. The ants typically nest in mounds constructed in open areas, pastures, and planted fields. The mounds dotting the land harden to dome-shaped structures that interfere with cultivation. Land where fire ants nest also becomes underused or useless owing to the risk of ant stings, and the ants themselves damage crops as diverse as potatoes, strawberries, and soybeans. They also collect around electrical wiring and equipment, apparently attracted by local electric fields. This tendency occasionally leads to grounded circuits and jammed contacts in air conditioners, pumps, and even traffic signals.

Despite years of effort, these aggravating pests have not yet been brought under control in the United States. They are much more successful here than in South America, where several dozen enemies have long held them in check. Lacking natural adversaries in the United States, fire ants have rapidly increased their numbers and continually conquered new territory. There are now five to ten times as many fire ants in their new home as in their native one. Completely eradicating them from millions of acres of land has not proved feasible, but scientists at the U.S. Department of Agriculture (USDA) recently began a program designed to bring them under biological control. USDA biologists have studied several of the fire ants' South American enemies, including some very small scuttle flies (*Pseudacteon litoralis* and several related species about 1 millimeter long) that lay their eggs in red imported fire ants. A field test now under way in Florida should determine whether scuttle flies can curb the ants here.

Pseudacteon scuttle flies belong to an absolutely intriguing group of parasitoids known as ant-decapitating flies, owing to their grotesque manner of destroying their hosts. If these flies succeed in controlling red imported fire ants, it will be largely through beheading them, one by one. Decapitation is only one of the remarkable events that commence with the tiny flies searching out foraging ants. USDA scientists believe that the flies locate fire ants by their scent, but they are still working to identify the active compounds. As the ants gather food, the flies hover and dart only a few millimeters above them. Both sexes of the flies are present, and they probably mate as they harass the ants. As soon as the ants become aware of the flies, they stop foraging and either retreat underground or curl individually into a C-shaped defensive posture. How the ants recognize the flies is still unknown. Because live immobilized flies do not alarm the ants, scent is probably not important to recognition here.

After some moments of darting and buzzing over an ant's head, an agile female fly swoops down and deposits a single egg in the ant's thorax (the second of its three major body segments, between its head and abdomen). Her assault requires only half a second, too fast to follow in detail, but it frequently knocks the ant off its feet. After a moment, the stricken ant rises up on its legs and stands motionless and still stunned for a minute or more. The ants possess little defense against the scuttle flies' forays. Guard ants take up a threatening stance, wagging their heads from side to side and waving their antennae, but the flies are too quick for effective retaliation. Very rarely a fly falls among the ants and is killed before it can escape. Their larval parasitism aside, the adult scuttle flies also debilitate fire ants by seriously disrupting their search for food. The ants return to normal foraging only twenty or thirty minutes after the flies depart.

A scuttle fly egg laid in a fire ant soon hatches to a very small larva or maggot, as a fly larva is usually called. There is too little room inside the ant's narrow thorax for the maggot to reach its full length of about 1.5 millimeters, and within three or four days it migrates forward into the ant's head. As it grows there over the next two weeks or more, the ant continues to behave normally, showing no outward sign of its fatal affliction. Then, as the maggot matures and prepares to pupate, enzymes under its control suddenly begin to dissolve ant membranes and tissues. Wiggling about within the ant's head, the maggot devours every bit of tissue it can reach. The ant's muscles, brain, and glands all disappear, and shortly nothing remains inside the ant's head but the maggot. Sometime during these singular events, the ant's head falls off, with the maggot safely inside.

The headless body is left standing upright and motionless, as the maggot industriously converts the severed head into its pupation chamber. Pushing aside the ant's mandibles and forcing open the mouth, the maggot positions its anterior segments to fill the mouth cavity. This anterior portion contracts and hardens into a darkened plate that completely blocks the opening. The plate is slightly convex and seals tightly, so the chamber wholly encloses the developing pupa, shielding it from attack by hostile ants. After about three weeks, a newly formed adult fly pushes aside the hardened plate and slips through the ant's mouth into the outer world.

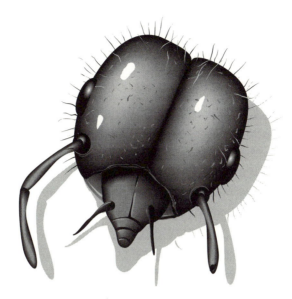

ILLUSTRATION 8 An ant-decapitating
fly's offspring has grown fat consuming the
entire contents of this ant's head.

Soon it will begin looking for fire ants and so carry its life cycle on
to the next generation.

Red imported fire ants precipitate these improbable activities
by initially drawing the flies to themselves through their own
chemicals, but the alluring compounds remain unidentified. The
only known attractants of ant-decapitating flies are those that lure
another small scuttle fly (*Apocephalus paraponerae*) to a giant tropical
ant named *Paraponera clavata*. These ants are Central America's
largest, and their venom is notorious for inflicting excruciating
pain. They are brown-black giants with soldiers 16–22 millimeters
(up to nine-tenths of an inch) in length, and even larger queens. The
tiny flies that parasitize these huge ants have habits unlike those of
the flies that assault red imported fire ants. They prefer to lay their
eggs in individuals that have been injured or killed while fighting.
Two compounds, 4-methyl-3-heptanone and 4-methyl-3-heptanol,
delivered from the ants' mandibular glands during combat provide

the signal that attracts the flies to the scene of battle. As the flies flit about, gravid females probe huge injured ants and then deposit their eggs. Once their egg laying is complete, male and female flies join together to feast on the dead and dying ants strewn across the battlefield. Greedily they drink up the casualties' body fluids, drinking until they are so bloated they can drink no more.

CHAPTER

Success through
Mimicry and Theft

6

Just as scuttle flies rely on chemical eavesdropping in exploiting red imported fire ants, tiny parasitic wasps that live among these ants also depend on chemicals for their survival. These wasps, an unidentified species in the genus *Orasema*, lay their eggs in the leaves, buds, and fruits of various plants regularly visited by fire ants. The eggs yield extremely small larvae, no longer than a small grain of salt (only one- to two-tenths of a millimeter), that attach themselves to ants foraging on the plant. The larvae may identify the ants by odor, and they may be so inconspicuous that the ants fail to notice them. In any event, when the ants return to their nest laden with food, the wasp larvae travel with them.

Once within the nest, the larvae make their way to the ants' brood chamber, where they take up residence posing as ant larvae. Their ruse is successful, and nurse ants care for them throughout their development, feeding and grooming them as though they were part of the ant brood. The wasp larvae repay the ants' generosity by eating their food and parasitizing ant larvae and pupae. The parasitized ant brood become malformed and die, but the wasp larvae prosper and then pupate in the ant nest. When an adult wasp finally emerges, it too is generally accepted by the ants. Should a curious ant examine it too closely, the wasp merely plays dead until the ant loses interest. Soon, the wasp scrambles to the nest entrance and flies away.

For this lifestyle to succeed, the wasp larvae obviously must avoid being recognized. If the ants were to detect them, they would suffer the fate of other intruders discovered in the nest: expulsion or death. Although some intruders escape detection in ant nests because they resemble their ant hosts, *Orasema* larvae, pupae, and adults all are easily distinguishable from their hosts, at least by the human eye. Fire ants, however, are much more concerned with smell than physical appearance. Fire-ant odor and local colony odor are what matter to them. By these scents, ants identify their own nestmates as distinct from other species and even from fire ants belonging to rival colonies. The wasps' strategy to avoid discovery is to smell just like fire ants. Larvae, pupae, and even adult wasps carry the distinctive scent of their particular fire ant colony. Shortly after the adult wasps leave the ant nest, this ant odor begins to fade.

How do these parasites acquire the protective scent that promises safety? There are two simple possibilities. One is that they steal scent chemicals from their ant hosts. A thin layer of the compounds they need coats everything throughout the nest, including, of course, all the ants. As nurse ants care for the immature wasps, the two species are continually in intimate contact, and the chemicals could pass directly from ant to wasp. The second possibility is that the wasps mimic the ant scent by synthesizing the appropriate chemicals and covering their bodies with them. Both of these approaches to chemical camouflage have precedents. Some nest parasites steal the scent they need, and others synthesize an imitation. It is not yet certain which the wasps do, but stealing the scent seems more probable. Reproducing the unique odor of a specific fire ant colony could be a severe challenge to the wasps. The particular colony odor each larva must match depends on its chance attachment to a specific ant just after hatching.

Like these wasps, many other living organisms steal or imitate a neighbor's chemicals, often to the detriment of others. The most straightforward chemical thieves appropriate desirable compounds directly by taking them in through their diet, avoiding the work of imitative synthesis. For a diminutive California limpet, this sort of appropriation is indispensable. Limpets are small marine mollusks with soft bodies covered by a single low rounded shell. They have a muscular foot they use to hold fast to a surface and to move about in their intertidal habitat, feeding at high tide on green plants, seaweed, and other algae. When the tide begins to ebb, limpets cling firmly to a rock or other surface, drawing their shell down tightly to resist being washed out to sea and to shield themselves from desiccation on exposure at low tide. Their way of life is ancient: Limpet-like creatures first appeared over half a billion years ago in the early to middle Cambrian period.

These particular California limpets are called *Tectura paleacea* and are adapted in shape and lifestyle to an uncommonly limited habitat. Most limpets live on rocks, some even maintaining a depression in a rock surface as a home to which they habitually return as the tide goes out. These *Tectura* limpets, however, spend their lives on blades of surfgrass (*Phyllospadix scouleri* and related species), grazing on the plant's surface layers. Surfgrass grows luxuriantly in the lower intertidal zone along the California coast, where it covers rocks with splashes of bright green and creates a safe haven for many small creatures. It is a member of the eel-grass family (Zosteraceae), which takes its name from its long narrow leaves. To accommodate themselves to life on the leaves, these tiny limpets have a parallel-sided shell, perhaps 6 millimeters long and 2 millimeters wide (one-fourth by less than one-tenth inch). These dimensions permit the shell to fit lengthwise on a blade of surfgrass quite precisely from one edge to the other. If disturbed, a limpet can clamp down snugly and remain immobile on its leaf.

A bed of surfgrass protects limpets and other small mollusks because large predators find the thin fluttering leaves relatively inaccessible. However, one local resident that preys persistently on these little creatures is the lovely six-rayed star (*Leptasterias hexactis*), a small pinkish starfish about 3 centimeters (1.25 inches) in diameter. Six-rayed stars hunt by moving along a surfgrass blade, waving their long, mobile tube feet here and there as they search out small prey to ensnare and stuff into their mouths. Unlike other mollusks in the surfgrass, the *Tectura* limpets show no unusual reaction to an approaching six-rayed star. They neither run nor fight, but simply pull their shell down on the leaf and remain motionless as a hunting starfish crawls over them Usually the starfish ignores them and continues its quest for food.

Starfish ignore the limpets because they fail to distinguish them from the background surfgrass. To the starfish's chemical sense, limpet and surfgrass are indistinguishable because both "smell" of chemical compounds called flavonoids. Surfgrass synthesizes these flavonoids, probably as a defense against herbivores, and limpets then ingest them as they nibble on the plant. Flavonoids do not repel the limpets, but become a crucial defense for them. The limpets incorporate flavonoids into their shells, but not their soft bodies, where they serve as a chemical disguise. A six-rayed star gliding along a blade of surfgrass detects flavonoids in both surfgrass and limpet shell and is unaware of the limpet's presence. The subterfuge is quite effective. *Tectura* limpets are only a minor component of six-rayed stars' diet despite the two animals' frequent encounters.

A number of other plant-eating creatures also turn compounds in their diet to advantage. One group of useful compounds known as furocoumarins are common defensive agents in plants as diverse as wild parsnip (*Pastinaca sativa*) and the common grapefruit (*Citrus paradise*). Many herbivores resistant to furocoumarins feed on these plants despite their defenses and devise good uses for the furocoumarins they ingest. One of the most novel of these is an application made unknowingly by humans. For years many physicians and their patients have realized that oral medications are sometimes more effective if taken with a glass of grapefruit juice, but no one knew why this was so. Investigators at the University of Michigan found the answer in 1997, when they reported that the active agent in grapefruit juice is the fruit's furocoumarins.

The investigators discovered that these chemicals inactivate a particular enzyme localized in the wall of the small intestine. This enzyme ordinarily destroys a portion of some drugs before they can move from the gut through the intestinal wall into the bloodstream. If the enzyme is inactivated, more of a drug may remain available to be absorbed and utilized. This effect of grapefruit juice is quite variable. The enzyme degrades only certain medications and its level differs widely from person to person. Also, newer preliminary evidence suggests that grapefruit juice has the opposite effect with some drugs and actually inhibits their absorption. For someone who generates a large amount of enzyme and takes a medication sensitive to enzymatic degradation, a glass of grapefruit juice can increase the drug's effective concentration considerably, even as much as ninefold. Without knowing it, people have long been making efficient use of furocoumarins "stolen" from grapefruit.

Exploiting furocoumarins may be useful, but it is scarcely a vital human activity. In contrast, some creatures must take chemicals from a neighbor to stay alive. One insect critically dependent on another species is a small fruit fly (*Drosophila pachea*) that we can call a senita fly. These flies live only in the Sonoran Desert of southern Arizona and northern Mexico in areas where senita cactus (*Lophocereus schottii*) grows. Senita is an impressive cactus that grows as a dense clump of erect green columns with spined ribs, and some varieties reach a height of 5–7 meters (15–20 feet). In the wild, senita flies breed successfully only in the rotting stems of this fleshy cactus. In the laboratory they survive only if their diet includes bits of the cactus's tissue. Without the cactus, the flies simply die. Among Sonoran fruit flies, their dependence on senita is unique. None of the desert's twenty or so other species of fruit flies has any interest in the cactus, for senita's tall columns are laced with toxins they are careful to avoid.

Unlike their relatives, senita flies are unaffected by these toxins. They flourish on an artificial diet containing even ten times as much toxin as the cactus produces. The plant's toxins fail to deter the flies, but the toxins themselves do not draw the flies to the plant. The flies come because their lives depend on a chemical requirement the cactus can satisfy. To understand this, we must note that all insects, along with innumerable other organisms (including humans), require cholesterol and cholesterol-related compounds in order to live. These compounds are collectively called steroids, and hundreds of them occur in nature and fill functions from cell-membrane components to sex hormones. Insects synthesize the particular steroids they need starting from cholesterol or other steroids ingested with their food. The great majority of insects are not very particular about the starting steroids for these syntheses. Their enzymes can make use of cholesterol as well as a variety of dietary plant and yeast steroids, converting any of them into the steroids they must have.

Although most insects are flexible about their steroid sources, senita flies definitely are not; They lack a set of enzymes indispensable for utilizing most steroids, cholesterol included. The enzymes they do have can handle only a few rare natural steroids, and only with these unusual compounds as starting points can they produce their own essential steroids. The single Sonoran source of these rare steroids seems to be senita cactus, and thus senita flies must incorporate the cactus into their diets or perish. As the only Sonoran fruit flies resistant to senita toxins, senita flies experience no competition from their relatives for the one plant that keeps them alive.

Historically, this lack of competition is probably responsible for the flies' total dependency on the cactus: The flies somehow lost the biochemical machinery needed for conversion of common plant steroids into those steroids they require. The loss went unnoticed, because the cactus provided an assured source of steroids the flies were still able to utilize. However, no other available plants provide these unusual steroids, so they must now consume senita cactus to survive.

Instead of putting their own dietary chemicals to use, some creatures rob others of particular objects or materials comprising chemicals they want. One striking instance of such robbery involves three kinds of insects: an aphid, a green lacewing, and an ant. To understand the crime, we must first see how these animals live and interact with one another.

Aphids constitute a group of about four thousand small insects with soft, pear-shaped bodies. They live on plants and feed on their juices, which they suck up in immoderate quantities. Their insatiable craving for plant juices is directly associated with a habit of utterly unrestrained reproduction: Aphids are exceedingly prolific and lay so many eggs that they require more nitrogen for synthesizing new proteins and nucleic acids than do most organisms their size. Consumed in large volume, plant sap can furnish sufficient nitrogen to meet their requirements, but in obtaining this nitrogen aphids also take in large amounts of carbohydrate. For many organisms this would be welcome nourishment, because carbohydrate is a convenient source of energy. The energetic needs of aphids, however, are rather limited. These are relatively sedentary animals, ordinarily content to remain on a plant, peacefully sipping its juices. This quiet lifestyle burns much less energy than does the active existence of honey bees or house flies. Consequently, aphids must periodically dispose of excess unwanted carbohydrate. They do this by excreting a sugary sweet fluid, known appropriately as honeydew.

Some aphids discharge honeydew as waste, leaving splotches on bushes and shrubs. Honeydew falling from trees can create a noticeably sticky mess on sidewalks and parked automobiles. In contrast to such wasteful habits, other more provident aphids turn honeydew into an asset, saving their honeydew and presenting it to certain ants in a mutually beneficial exchange. An ant approaches an aphid and signals it by tapping or stroking. The aphid obligingly secretes a drop of honeydew, which the ant promptly consumes. Honeydew is a significant source of carbohydrate for these ants, and in return for it they effectively adopt the aphids, providing them with constant protection from their adversaries. The aphids badly need the ants' services. Their juice-filled bodies and slow ways make them attractive targets for parasitoid wasps, as well as predatory beetles and insect larvae. Undefended, aphids' only recourse is rapid reproduction, but watchful ants can drive off or destroy would-be assailants. Some ants watch over entire flocks of aphids, solicitously herding them from place to place on a plant for better nourishment, and so more honeydew for themselves.

Two insects in the northeastern United States that engage in exchange of honeydew for protection are woolly alder aphids (*Paraprociphilus tessellatus*) and black carpenter ants (*Camponotus pennsylvanicus*). Groups of several hundred aphids guarded by dozens of ants range over branches and leaves of speckled alders (*Alnus rugosa*). The aphids stand out against the dark alder branches as small woolly globs, because they cover themselves with a fluffy white secretion that is largely a single waxy ester. Vigilant black ants patrol the aphids, "milking" them regularly and defending them from attack. The ants are effective, but despite their diligence, one predator habitually eludes them and reaches the aphid flocks. This is the larvae of a green lacewing (*Chrysopa slossonae*), whose diet consists exclusively of woolly alder aphids. An entire family (Chrysopidae) of adult insects with long antennae and glossy wings is known as green lacewings. All enjoy a reputation as gardeners' friends, because as both larvae and adults they prey ravenously on aphids and other soft-bodied insects. They pierce their victims' bodies and suck them dry, to leave behind only a shriveled husk.

Lacewing larvae that feed solely on woolly alder aphids must somehow avoid attracting the attention of guard ants if they are to survive. To escape notice, they camouflage themselves with woolly wax taken from the aphids. A larva comes upon an aphid and shoves its head deep into the aphid's wool. Working its mandibles as a fork, it pulls away a bit of wool and, arching its head, transfers the wool onto its own back. Here the wool becomes entangled and held in place on stiff hooked bristles. Plucking wool from several nearby aphids, a larva can complete its masquerade within twenty minutes. Once it is covered with wool, it is safe among the aphids. Disguised larvae are much the same size and shape as aphids,

ILLUSTRATION 9 To complete their life
cycle woolly alder aphids must feed on
maple leaves in the spring and then move to
alder in the summer.

making it virtually impossible for a human to visually distinguish
the two.

The ants also fail to discern the wool-covered larvae among
their aphids. When an ant meets a disguised larva, it touches the
insect's woolly coat with its antennae. Then it usually goes on about
its business, its chemical senses satisfied that it has encountered an
aphid. Occasionally, an ant investigates further by biting into the
wool. After a single bite, it backs away. Carefully cleaning the wax
from its mouthparts, it scrutinizes the larva no further. To the ants,
the woolly wax evidently signifies *aphid*.

Almost perfectly, concealed, disguised larvae are free to glut
themselves on juicy aphids. Undisguised larvae receive quite different treatment, of course, for the ants immediately perceive them
as a menace. If a naked larva is lucky, the ants only throw it out of
the alder. Aggressive ants may bite larvae or even kill them. A

larva that survives an ant assault uninjured retreats to a secluded part of the aphid colony to renew its camouflage.

A different reason for appropriating a material object comes from the kauri (*Agathis alba*) and merkus (*Pinus merkusii*) pine forests of Malaysia. Local people collect soft resin from these trees and sell it as a cash crop. Under the name Malay damar, this exudate is exported as a valued ingredient for high quality varnishes.

Among the uncountable insects inhabiting the damar forests, several occur that also value this resin. These are carnivorous assassin bugs (*Amulius* and *Ectinoderus* species) that feed on tiny stingless bees and other small insects. They collect damar and smear it on their front legs. After coating its legs, an assassin bug takes up a position on a tree trunk to await its prey, with its head downward and damar-covered forelegs extended to form a sticky trap. If a careless insect stumbles into the trap and gets stuck, it becomes the assassin bug's next meal.

Do assassin bugs really collect damar for its chemicals? The resin serves their needs because it does not harden quickly but remains soft and sticky. Softness and stickiness are physical properties, but they are also direct consequences of the resin's chemical composition. It seems fair to credit the bugs' employment of damar to its complex mix of chemicals. We could also ask whether green lacewing larvae appropriate aphid wool for its chemicals. In that case, the significance of chemical properties appears unequivocal, as the ants' chemical sense evidently equates the waxy wool with aphids.

The ultimate examples of usurping physical objects come from creatures that appropriate whole organisms for their chemical content. Certain spiders, for example, treasure the dead bodies of ants they have fed upon. Ants have relatively few predators despite their great abundance and accessibility. Their biting mandibles, painful stings, and communal way of life make them troublesome prey, and most predators can secure a meal elsewhere with less effort. One animal that does prey on ants is a tiny South American crab spider (*Strophius nigricans*) with a technique for hunting ants that also furnishes protection from its own enemies. These black spiders are only 4–5 millimeters across (one-fifth of an inch), the same size and color as the carpenter ants (*Camponotus crassus*) that are their prey. The spider's trick is to carry a dead carpenter ant aloft as it moves about, walking in a zigzag fashion that closely simulates an ant bearing a dead nestmate. Carrying its trophy, the spider apparently looks and smells like an ant. The visual and chemical mimicry probably protects it from wasps and lizards that might pounce on a spider but would forego attacking an ant.

The trick also affords the spiders the means of capturing their prey. A spider approaches its ant prey holding a dead ant before it as a shield. On seeing the corpse and recognizing its odor, an ant occasionally advances for closer inspection. When the ant is very near, the spider drops the corpse at its feet and runs around behind the ant. The spider grabs the ant from the rear, quickly bites it, and injects a potent venom. The ant struggles to escape, fiercely biting the dead ant in its agitation. The spider retracts its first pair of legs as far as possible, carefully keeping them well away from the struggling ant's sharp mandibles. As the venom takes effect, the ant weakens and finally drops the corpse. The spider then lifts the ant high in the air and repeatedly taps the ant's antennae and legs with its own front legs. When the ant no longer responds to this probing, the spider spins it around and grips it by the neck. Without mutilating or crushing its body, the spider pierces the ant's hard outer skeleton and drains it of its contents. After this meal, the fresh corpse will be the spider's shield until it feeds again.

Rather than carry a corpse, a quite different animal kidnaps a living organism as its chemical defense. This is the strategy of a tiny marine creature called an amphipod, a member of a widespread order (Amphipoda) of small crustaceans. The tens of thousands of their species represent an assortment of types and lifestyles, many looking like tiny shrimp, to which they are related. Marine amphipods are one of the invertebrate success stories and can be found in almost all saltwater environments, feeding on plants, animals, or detritus. Many are microscopic or nearly so. Some are at home in the open sea as part of the living soup called plankton, while others burrow into sand or dwell hidden among seaweeds. Still others build themselves tubes affixed to rocks or attach themselves to a jellyfish or tunicate.

One amphipod that lives as plankton in the frigid antarctic is a creature called *Hyperiella dilatata*. This little creature is about 2 millimeters long and has the ill-fated distinction of being a favorite food for several common plankton-eating antarctic fishes. Unlike some of its relatives, *Hyperiella* has no spines or other physical features that might discourage predatory fish, nor does it possess any compounds suitable for chemical defense.

Nevertheless, *Hyperiella* is not defenseless. It has developed an amazing way of living safely among its enemies. *Hyperiella* arms itself against predators by capturing and carrying around wherever it goes a small mollusk called *Clione antarctica* that can be its own size or larger. This is a heavy load to bear, but for *Hyperiella* this constant burden is a matter of life or death. As long as the amphipod retains its captive, fish leave it alone. Some inconvenience and loss of mobility are a small price to pay to escape being eaten.

Clione belongs to a group popularly known as sea butterflies and is related to snails and sea hares. Its odd name refers to two winglike flaps that sea butterflies beat in swimming through the water. Compared to some other species, *Clione* is a rather weak swimmer and beats its flaps no more than twice a second. During the spring and summer months, these animals are very abundant and swim in conspicuous swarms along the antarctic continental shelf. They have no protective shell and appear in general to be poorly defended creatures.

Despite the easy availability of *Clione*, plankton-eating fish leave them alone, even fish that have a taste for other sea butterflies. It sounds as though *Clione* has a strong chemical defense, like other shell-less mollusks, and this is correct. Its tissues contain a compound called pteroenone that is a feeding deterrent. Fish will not consume pteroenone, and the few that sample these mollusks reject them at once. *Clione* appears to synthesize pteroenone for it-

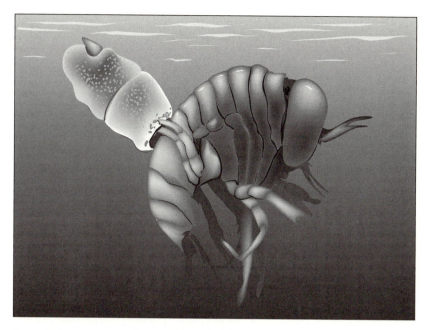

ILLUSTRATION 10 An amphipod holds a
small sea butterfly in place with two of its
pairs of legs. Although these sea butterflies
grow much larger, amphipods prefer to carry
small ones.

self rather than to acquire the compound from its diet. It is a specialist carnivore, feeding regularly only on one or two mollusks
that contain no trace of pteroenone.

The amphipod *Hyperiella* takes advantage of *Clione*'s chemical
defense. Fish that mistakenly snap up a *Hyperiella* equipped with a
sea butterfly immediately spit the pair out. In a laboratory test, fish
ate all the naked *Hyperiella* presented to them, and of a dozen
armed amphipods, they refused ten. In the two cases where they
consumed an amphipod initially carrying *Clione*, the pair came
apart in the fish's mouth. The fish swallowed the amphipod but rejected the sea butterfly. Apparently, the amphipod holds its body-
guard tenaciously, or the two would separate more often.

Hyperiella is not the only amphipod to borrow another organism for defense. Far from Antarctica in the Caribbean Sea, there is

another amphipod, *Pseudamphithoides incurvaria* (PI), with a comparable chemical defense. However, PI does not appropriate an entire organism, but only small bits of one. Its defense lies somewhere between *Hyperiella*'s kidnapping a living sea butterfly and lacewing larvae's removing woolly wax from their aphid prey.

PI originally attracted attention owing to two unusual habits. It feeds on seaweed, and although many seaweed-eating amphipods have broader tastes, PI nibbles at only one species. It feeds only on a flat-bladed brown seaweed called *Dictyota bartayresii.* Specialist feeders are less common among marine organisms than on land, but PI seeks out this one species even when other *Dictyota* seaweeds are more abundant.

Dictyota species are rich with unpleasant-tasting chemicals to discourage grazing fishes. Amphipods and other small creatures often find safety in among such unpalatable seaweeds, where fishes are infrequent visitors. The chemicals in *Dictyota bartayresii* are not unpleasant to PI, and in fact the amphipod uses them to identify the seaweed it eats. Conceivably, PI could also sequester these distasteful compounds for its own protection; perhaps surprisingly, it does not.

Instead of sequestering the seaweed's compounds as a defense, PI appropriates the seaweed itself. The amphipod constructs a millimeter-sized domicile, joining together little bits of seaweed to fashion a structure something like the shell of a clam, with the two halves hinged by a threadlike secretion. PI is 1–2 millimeters long and fits nicely inside this seaweed structure, with its head and several forward pairs of legs sticking out so that it can swim. In this way, the amphipod remains mobile while safe and secure within its seaweed home. This is an effective defense. In a laboratory experiment, fish quickly snapped up naked defenseless amphipods but rejected those inside their domiciles. Just as with *Hyperiella*, if a fish took a protected PI in its mouth, it quickly spat out the amphipod and its domicile. As protected amphipods left the fish's mouth, some unfortunate ones fell out of their domicile, whereupon the fish immediately pounced and devoured them.

Rather than steal chemical compounds, some creatures produce particular chemicals that serve another species as a pheromone, finding advantage in mimicking the other species' signal. (Sometimes the mimicking creature may have possessed the chemicals before the one being mimicked, originally employing them for some other purpose and later adapting them to serve as a counterfeit pheromone.) Some slave-making ants, for example, disseminate their ant victims' alarm pheromone during their raids, inducing adults to flee and abandon their brood to lifelong slavery. Certain spiders make and release the sex pheromones that receptive female insects broadcast to lure prospective mates from afar. Ardent males hasten to the attractant expecting to discover a responsive mate, only to encounter a predacious spider instead.

Some parasitic bee species imitate one of their host's signals in laying their eggs. Here, the female parasitic bees must get this signal from the male bees, which are responsible for synthesizing and storing it. To see how this works, we must briefly examine both these parasites and their hosts. They belong to different families, but both parasites and hosts are members of a large group known as solitary bees. Although we are accustomed to thinking of honey bees and their highly complex communities as representative of the way bees live, the great majority of bees follow a solitary rather than a social existence. For solitary bees, there are no hives filled with workers, nurses, and queens. Each female lays her own eggs and makes provision for her offspring in the fashion of her particular species.

One family of these solitary bees consists of smallish wasplike creatures commonly called cuckoo bees. As their name implies, they are parasites that lay their eggs in other bees' nests, thereby evading all the responsibilities of parenthood. There are several similar species of cuckoo bees, but those we shall focus on are called *Nomada marshamella*, and the solitary bees they parasitize are *Andrena carantonica*. We can designate them parasite and host, respectively. Female host bees nest in the ground, each one digging a main burrow with short lateral tunnels. At the end of each tunnel she constructs a nest cell and waterproofs its walls with a secretion from her abdominal gland. This secretion is largely an oily substance known to chemists as farnesyl hexanoate. It is a component of host nest-odor that we shall abbreviate as F6. After stocking a cell with pollen and nectar from nearby flowers, a host bee lays an egg on top of this food supply and seals the cell closed. When the egg hatches, the larva will find ample nourishment in its private chamber.

In contrast to the host's careful preparations, the parasitic females neither build nests nor gather food for their progeny. Instead, they discreetly deposit their eggs in partially finished host nests that they locate through chemical and visual cues. If two parasitic females should converge on the same nest, they fight vigorously for the right to exploit it, butting heads, pouncing on each other, and releasing chemicals from their mandibular glands. Eventually one bee prevails, and after chasing her competitor away she stations herself at the mouth of the nest. As soon as the busy nest-builder leaves on a foraging flight, the parasite enters the nest to look for partially provisioned cells still under construction. When she finds one, she lays an egg in it and departs. The host bee returns to complete the cell, then lays her own egg and seals the cell. The parasite's egg hatches first and the parasitic larva consumes the host's egg and food supply in fueling its own development. One more parasitic bee is launched at its host's expense.

Though two parasitic females struggle strenuously over a nest, the inevitable encounters between parasite and host at the nest are unexpectedly peaceful. There is no fighting because the host accepts the parasite as one of her own kind. This cannot be a visual error, since the two bees show little general resemblance. The explanation is more intriguing and can be traced to the parasites' synthesis of F6, the host bees' nest-odor and waterproofing compound. Surprisingly, the male parasites make and store F6 in their mandibular glands, but F6 is not in female parasites' chemical repertory. This seems strange: Females produce no F6, although with it they could mimic host nest-odor and so promote their own acceptance.

The arrangement remains puzzling, but the bees easily resolve the problem (or what we see as a problem). When the parasitic bees mate, the male simply sprays some of his F6 on the female. The compound is relatively involatile and so persists on the gravid female as she seeks host nests for her eggs. Perfumed with F6 from her mate, the parasitic female penetrates the host nest and completes her task unchallenged. At least three other species of parasitic *Nomada* bees gain acceptance by their *Andrena* hosts in this way.

Mimicry of pheromones is also basic to a bargain between certain caterpillars and ants. Most caterpillars live in fear of carnivorous ants, which see these slow-moving, soft-bodied larvae as easy prey. As early as the eighteenth century, however, naturalists were fascinated to discover that some caterpillars engage ants as their protectors. We now know that these caterpillars all belong to two closely related families of butterflies (Lycaenidae and Riodinidae) that include species commonly known as hairstreaks, blues, and metalmarks. Half or more of the caterpillars in these two families have guardian-ants, and the majority of them are able to associate with more than one species of ant. Some of them absolutely require the ants' protection, whereas others can survive without it. Although aphids offer only honeydew, caterpillars contribute two or three different chemical secretions as compensation to their guardians. Interactions between several different genera of ants and caterpillars from these two families differ considerably in detail, and rather than explore a specific example, it is suitable to discuss them in general terms.

On encountering a caterpillar, a foraging ant must decide whether or not to attack. An ant-friendly caterpillar attempts to influence the ant's decision by secreting a liquid from glands in its skin. An ant willing to negotiate with caterpillars explores the secretion with its antennae and presses its mouthparts against the glands, apparently collecting the liquid. Some caterpillars offer a nutritious secretion of amino acids. Others release a good imitation of the ants' brood pheromone, which is the scent characterizing the ants' larvae and eggs. Whatever its nature, the secretion effectively appeases the ant. Instead of attacking, it taps its antennae around an organ on the caterpillar's back. On this signal, the organ delivers droplets of a nectarlike solution. The ant sips the nectar and may share it with nearby nestmates. This energy-rich solution is the caterpillars' principal offering to the ants, playing a role analogous to aphids' honeydew. In return for it, the ants guard the caterpillars just as they do aphids.

In addition to nectar, many caterpillars dispense another secretion to influence the ants' behavior in the face of danger. When these caterpillars are disturbed or under attack, another secretory organ that lies just behind the nectar organ extrudes small tentacle-like structures. These structures broadcast a volatile signal that instantly animates nearby ants. The ants become sharply more alert, excitedly running around and over the caterpillar. The signal is the ants' alarm pheromone, and it readies them for imminent conflict. Along with this pheromone, the caterpillar may also deliver drops of nectar to reward the ants for their increased alertness in a crisis.

Field experiments have shown that ants indeed defend their caterpillars effectively. Among the caterpillars' most persistent foes are marauding social wasps. Although adult wasps eat only plant products, they spend much of their time hunting caterpillars to kill, cut up, and feed to their carnivorous larvae. Caterpillars guarded by ants withstand wasp attacks much better than those without protection.

The caterpillars' enhanced survival implies that their imitation signals and reward secretions are worth the energy expended on their synthesis and dissemination. On the other hand, it is questionable whether ants really profit from their deal with the caterpillars. Had they eaten the caterpillar forthwith on their first encounter, the ants would have gained about the same number of calories as from milking it over its lifetime. However, the situation is complex, and only more information can settle some issues here. One possibility is that the nectar affords compounds the ants must have in their diet but would not obtain by consuming the caterpillars themselves. Another is that caring for caterpillars rather than killing them may lead to an increased number of caterpillars in the future, and that this is to the ants' ultimate advantage.

Finally, there is at least one parasite that practices deceit in a negative sense. That is, as part of a scheme to deceive its host, it ceases to elaborate a family of chemicals that would reveal its presence. This unusual deception involves two closely related wasps that live high in the French Alps. Both the parasite (*Polistes atrimandibularis*) and its host (*Polistes biglumis bimaculatus*) are social wasps belonging to a group called paper wasps. These wasps build papery nests consisting of a single layer of exposed, downward-pointing cells suspended by a narrow stalk. Together, the nest and stalk resemble a rather flat umbrella held upside down. Most *Polistes* construct such nests, but the parasitic *Polistes* are incapable of building nests for themselves. Moreover, they possess no worker caste and so are unskilled in the housekeeping chores necessary for sustaining a nest. They subsist only by commandeering nests built by a host queen and maintained by her nestmates.

A parasitic queen begins her search for a suitable egg site early in July, about a month after a host queen has founded a new nest. When the parasite locates a host nest, she simply moves in. At this time parasite and host carry quite different recognition compounds on their bodies and so are readily distinguishable. The parasitic queen's recognition mixture is rich in compounds called alkenes, but these chemicals are totally absent from the host queen's odor. On installing herself in the nest, the parasite is initially unassertive, but as she begins to lay her eggs, she becomes increasingly dominant. In contrast, the host queen starts out aggressively resisting invasion of her nest, but she gradually appears more at ease and reconciled to the parasite's presence. Once the host is calm, the parasite has no difficulty in taking over the entire nest.

The progressive decline in the host queen's aggressivity reflects chemical changes in her parasitic intruder. As the parasite takes command, her surface chemicals come to resemble her host's more and more. The parasitic queen either simply acquires the host chemicals by direct contact or else synthesizes them herself—the same two possibilities available to the *Orasema* wasps discussed earlier. More important, she also effects a chemical alteration that is truly remarkable. As the parasitic queen takes on her host's odor, her own telltale alkenes begin to vanish. She is somehow able to suspend synthesis of these signature compounds that are foreign to her host. By the end of July her alkenes have completely disappeared, and the two queens have indistinguishable odors. No longer able to identify the egg-laying guest as an outsider, the host queen is indifferent to her presence.

During August, the host queen typically abandons her nest, but the parasite stays on. By the end of August, a new generation of parasites emerges and the parasitic queen has no further need for her disguise. Her alkenes reappear, and with them she resumes her own unique odor. How she turns the synthesis of alkenes off and on is still a mystery.

Bacteria: Chemical Complexities in Simple Cells

7

Bacteria are among the simplest living organisms, each individual consisting of a single cell that carries out all the necessary functions of life. Like other cells, a bacterium reproduces by dividing in two, metabolizes food to obtain energy, and discards its wastes. Bacteria are typically smaller than the cells of multicellular organisms and, unlike them, lack a nucleus to enclose their DNA. Even these simple organisms have a chemical sense, and as they move about in their environment they respond to chemical nutrients or toxins by moving toward or away from them, respectively.

Many bacteria perform services we find useful and even necessary, although we rarely take note of them. Some bacteria participate in the decomposition of dead plants and animals and so help recycle chemicals that are otherwise locked away from the living world. Some are sources of antibiotics that have revolutionized the treatment of infectious disease in the past fifty years. Others are responsible for nitrogen fixation, converting relatively inert atmospheric nitrogen into biologically useful ammonia—a conversion that is not simple in the chemical laboratory. Still other bacteria are indispensable in the world's kitchens, for without them there would be no cheese or yogurt, nor any of hundreds of other traditional milk products. We could not enjoy pickles or sauerkraut, and there would be no vinegar to dress salads. A few other bacteria are the workhorses of modern biotechnology. Using genetic engineering techniques that transfer genes from one organism into another, biologists convert these bacteria into living factories for commercial production of insulin, human growth hormone, and other medically valuable proteins.

Most of the five thousand or so known bacteria are innocuous and arouse little general interest. A notable exception is the group of pathogenic organisms responsible for infectious diseases in humans, animals, and plants. Since 1876, when Robert Koch demonstrated that anthrax results from infection by a bacterium (*Bacillus anthracis*), these pathogens have been viewed as enemies of humanity. Before the advent of antibiotics, "germs" were a major cause of death throughout the world, as they still are today in those regions where pure water is a luxury, proper sanitation rare, and antibiotics unavailable. Tainted food and water favor such pathogens as *Vibrio cholerae*, which causes cholera, and several *Shigella* species, which lead to shigellosis, or bacillary dysentery. These two diseases alone are responsible for hundreds of thousands of avoidable deaths each year. Unfortunately, there are much worse bacterial killers. Tuberculosis alone claims two to three million lives annually, and WHO has anticipated a higher annual toll in recent years than ever before.

In the more fortunate developed world, interest in bacterial pathogens declined in the mid-twentieth century as antibiotics became common and the incidence of life-threatening infectious disease dropped dramatically. However, several independent events over the past thirty years have refocused serious attention on pathogens. New bacterial diseases emerged in the mid-1970s: Toxic-shock syndrome, Lyme disease, and legionnaires' disease were all novel afflictions whose names quickly became familiar. Medical scientists soon found that Lyme disease and legionnaires' disease result from previously undescribed organisms, making them seem even more threatening. More recently, the bacterial origin of peptic ulcers has become widely recognized, and in mid-1998 evidence first appeared suggesting that a very small bacterium new to sci- ence may be to blame for kidney stones. In addition, deaths in the United States from infectious diseases (excluding virus-caused ac- quired immune deficiency syndrome, or AIDS) rose 22 percent in the decade following 1982. Beyond these unsettling events, the most significant factor rekindling interest in pathogens has been the growing number of antibiotic-resistant bacteria. Some particu- larly adaptable pathogens develop resistance to new antibiotics al- most as fast as the drugs become available. During 1995 far more people were killed in New York City by antibiotic-resistant staphy- lococcus infections than were murdered. Bacterial resistance to an- tibiotics is a severe and growing medical problem and the subject of much current research. It is by no means a simple issue, but one significant contributing factor is misuse and overuse of antibiotics by both physicians and patients. A National Institutes of Health (NIH) study released in May 1998 reported that 20–50 percent of the drugs prescribed by physicians to their office patients were unnecessary.

All this suggests that devoting some attention to bacteria is well worthwhile. These tiny organisms may be relatively simple, but they vary greatly in metabolism and physiology from one species to another and exhibit a remarkable capacity for evolutionary change when confronted with environmental challenges. Further- more, bacteria employ special chemicals much as multicellular creatures do. They interact chemically with other kinds of organ- isms, and individual bacteria of some, or perhaps many, species communicate with one another. Chemically, bacteria prove to be very adept.

Just as we found that various larger organisms take other crea- tures' chemical compounds for their own use, there are also chemi- cal thieves among bacteria. One of the shigellosis-causing patho- gens (*Shigella flexneri*) spreads so quickly that the host's immune system has little chance of repulsing its invasion. *Shigella* uses a clever trick in executing this microbial blitzkrieg. After ingestion, each bacterium passes from the host's gut into a cell in the intestinal

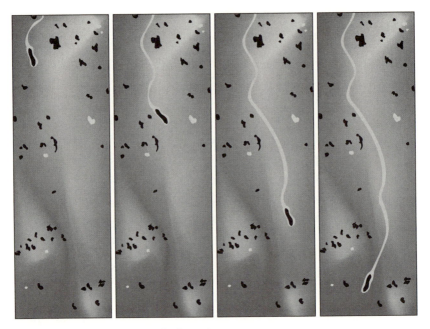

ILLUSTRATION 11 *Shigella flexneri*
sweeps through cytoplasm propelled by its
cometlike actin tail.

wall where it is concealed from the host's immune system. Protected and shielded from detection, it undergoes rapid prolifera-
tion. Each *Shigella* cell then appropriates a bit of a host cell protein
called actin. The molecules of this common structural protein form
rodlike filaments that lend form and rigidity to intestinal cells, just
as they do to other cells throughout much of the biological world.
A *Shigella* cell steals some of these actin filaments, and with the par-
ticipation of a protein of its own fashions itself a tail by affixing one
end of the filaments to its own surface. Propelled by complex chemical activity in its new tail, *Shigella* speeds through its host cell and
dashes quickly from one host cell to another, proliferating and
spreading infection as it goes. The host's immune system seldom
locates these swift marauders.

There is another dangerous bacterium unrelated to *Shigella flexneri* that makes similar actin tails for itself. This is *Listeria monocyto-*

genes, a food-transmitted human pathogen that was responsible for a serious outbreak of disease in late 1998. *Listeria* can bring about severe meningitis and infrequently induces miscarriages. It turns up in processed meats and milk products, occasionally contaminating the cheese in commercial pizza. In turning actin filaments into tails, both *Shigella* and *Listeria* make use of only a single bacterial protein of their own. The proteins from the two microbes are structurally quite dissimilar and so do not share a common origin. Apparently, bacteria have independently evolved mechanisms for making actin tails at least twice.

Chemical thievery may also operate in the opposite direction, when microbes supply rather than secure chemical compounds. The organisms taking microbial chemicals are typically multicellular and harbor parasitic bacteria or have bacteria somewhere in their food chain. These contacts permit the larger organisms to appropriate the microbes' chemicals for their own use. One creature of this sort is the dragon fish (*Malacosteus niger*), a sharp-toothed predator of the deep sea with an ingenious hunting technique built around bacterial chemicals. Most deep-sea fish are visually most sensitive to the blue light that penetrates the ocean depths farther than other wavelengths. Moreover, many inhabitants of this dim world are bioluminescent: they emit biologically generated light that results from certain chemical reactions. In most cases their light is blue.

Using this blue light to search the dark waters for prey would present an obvious disadvantage: A predator with a light other creatures can plainly see will very likely reveal its own presence long before sighting its prey. The dragon fish avoids this problem and remains concealed by employing a red light in searching for food. Since red light is invisible to other deep-sea animals, it gives the dragon fish's prey no advance warning. They never realize a predator has spotted them until it is too late. For this scheme to work, the dragon fish must of course be able to see whatever its red bioluminescence illuminates. Unlike other fishes, it has to see red. There are several ways to do this, but how the dragon fish actually sees red remained an unsolved puzzle until mid-1998. It is now clear that the dragon fish's eye contains not only the common blue-sensitive pigments but also a mixture of red-sensitive pigments absent in other fishes. Red light striking the molecules of these unique pigments excites them energetically. The excited molecules quickly shed their excess energy by passing it on to the ordinary blue-sensitive pigments, exciting them in turn. This secondhand excitation is essentially the same as the excitation the ordinary pigments experience when struck by blue light. Because the excitation appears normal, the blue-sensitive pigments respond in their normal fashion, which is to send signals along the optic nerve to the fish's brain. In this way, the brain receives the same signals from both blue and red light, and thus the dragon fish perceives both blue and red.

The red-sensitive pigments are a nice solution to the fish's problem, but what are they and where did they come from? On analyzing them, chemists were perplexed to find they were chlorophyll-like compounds. This was peculiar, because elsewhere chlorophylls are linked to photosynthesis. Green plants and photosynthesizing bacteria produce chlorophylls, but fish do not. The explanation came when chemists recognized that the fish's pigments were closely related to bacterial chlorophylls present in so-called green sulfur bacteria. These are *Chlorobium* species that inhabit waters rich in sulfides. They are much too small to whet a dragon fish's appetite, but microscopic planktonic organisms eat them in vast quantities. The plankton are themselves consumed equally voraciously by tiny crustaceans called copepods, which in turn are the prey of larger crustaceans that dragon fish do consume. It appears that each of these organisms sequesters the bacterial chlorophylls, and that the dragon fish secures its red-sensitive pigments indirectly from these bacteria that lie at the bottom of its food chain.

A group of animals that is far simpler than dragon fish and is totally dependent on bacteria for its lifestyle and livelihood includes two families (Steinernematidae and Heterorhabditae) of nematodes, or roundworms. Here it is not a matter of worms exploiting bacteria but rather of worms and bacteria combining their chemical resources in a joint enterprise advantageous to both. These nematodes are slender worms about one-half to 1 millimeter long. Though tiny from our point of view, they are nonetheless multicellular animals with diversified body parts. They are much more simply constituted than insects or mollusks but also are much larger and more complex than microbes. The bacteria cooperating with these worms are several species of rod-shaped cells about five-thousandths of a millimeter long belonging to the genus *Xenorhabdus* or *Photorhabdus*. Each species of worm has one species of these bacteria residing in its gut. The two creatures are never far apart: The worms carry the bacteria from an early age and the bacteria exist only in the company of the worms. Living and working together, the two have flourished and enjoy a successful way of life.

The nematodes are themselves insect parasitoids that are not very particular about their hosts. Fly maggots, moth larvae and pupae, beetle larvae, and numerous other hosts are all acceptable to them. Some of these nematodes do have narrow preferences, but one widespread species invades more than two hundred different kinds of insects. Juvenile nematodes infected with their bacteria seek out a host to parasitize, typically gaining entry through one of its body orifices. Some species enter through a hole they scrape in the insect's cuticle using a "tooth" on their head. Once inside the insect, the worms force their way through soft tissues and into their host's central body cavity.

Here deep within their host, the nematodes discharge their bacteria. The microbes go to work immediately, synthesizing and secreting a variety of chemical compounds that permit the invaders to take over their host. The most important of these is a toxin that kills the insect over the next day or two. As soon as it is dead, its cadaver affords a feast and a home for both bacteria and nematodes. The growing worms feed indiscriminately, devouring many of their bacterial partners along with insect tissues. While the parasitoids feed, the dead insect remains surprisingly fresh. The bacteria provide antibiotics to keep putrefying microbes at bay.

The nematodes in turn contribute compounds that protect the bacteria. An insect's immune system can make proteins to fight off invading microbes. If these proteins annihilate the bacteria before they release a lethal dose of toxin, the invaders' effort will fail. Chemicals from the nematodes block this defense, leaving the insect vulnerable to poisoning. Between them, then, worms and bacteria reduce the insect to a rich food source. Their prize is also an ideal site for the nematodes' reproduction. Two or three weeks after their initial attack, the nematodes have matured and mated. Their next generation is quickly under way. Soon, up to a half-million new juvenile nematodes are ready to leave the cadaver and make their way in the world. At the same time, the microbes have also multiplied and greatly increased their numbers. The young nematodes infect themselves with their bacterial partners, and together the two set forth to hunt for new hosts.

Until twenty-five or thirty years ago, few laboratories around the world were seriously interested in these insect-eating nematodes and their bacteria, but agricultural and industrial scientists are now studying them enthusiastically. Mounting pressures to limit the use of synthetic chemical pesticides have had their effect, and the nematode-bacteria complex may offer a natural environmentally safe biopesticide. Both creatures are harmless to vertebrates, but the bacterial toxins are fatal to a wide spectrum of agricultural pests. An added advantage is that the nematodes actively seek out their hosts, something other pesticides cannot do.

The simplest way to employ these natural insect-killers would be to treat crops directly with infected nematodes, since they can be raised, stored, and spread relatively cheaply. In fact, this idea was first tested in 1932, long before there was any idea how the nematodes destroyed their hosts. The worms showed considerable promise in these early trials, but the project ultimately collapsed with the advent of the Second World War and the subsequent introduction of synthetic pesticides. In more recent field tests, infected worms have scored up to 100 percent pest mortality in a few days.

Turning these creatures into a practical pesticide is beginning to look encouraging.

A popular belief is that each bacterium is a simple isolated cell, but the fact is that bacteria secrete pheromones for communicating with their own kind. In a few cases these messages coordinate extraordinary collective behavior resembling that of a multicellular organism. Discovered only in the past decade, bacterial cell-to-cell communication has become an area of vigorous research, owing in part to its potential implications for understanding vertebrate development. In the complex transformation of a fertilized egg into a new individual, whether frog, chicken, or human, signals from one embryonic cell to another control the course of many events. Studying bacterial communication may furnish useful insights into these developmental signals.

What could one bacterium possibly have to say to another? Quite a number of things, it turns out, the most elementary one being, effectively, "I am here." Pooling this information from many cells, a group of bacteria can determine whether their total number is sufficient for some collective action. A pretty example comes from a bioluminescent marine bacterium called *Vibrio fischeri*. More than two dozen species of squid gather these microbes as a source of light. The squid have translucent superficial pockets called light organs, which they fill with bacteria collected from the sea. Within the light organs, the bacteria find plentiful nutrients and a comfortable, protected environment. In return for this safe habitat, they emit light beneficial to the squid.

These same bacteria also live free in the ocean, but there they produce no light. When grown in laboratory culture, they are bioluminescent only when their population density becomes very high. It is easy to see that luminescent bacteria living in the ocean would simply advertise their presence to predators. In the open sea, keeping the light off is safer. However, bacteria that are secure in a squid's light organ can safely luminesce to earn their keep.

How do the bacteria "know" when to turn their bioluminescence on or off? It may look as though they decide shrewdly when to shine, but actually a single pheromone acts as a simple switch to control their luminescence. Each bacterial cell secretes a chemical signal whose message is, "Count me, too; I am here." When the bacteria are dispersed in the open sea, this signal has a very low concentration. After a squid takes them up, however, the pheromones from innumerable cells accumulate in a closed light organ, where they quickly build to a high concentration. The bacteria sense the signal in their environment, and when the concentration attains a certain threshold they begin to emit light. This mechanism ensures that the bacteria become bioluminescent only when a large number are present in a confined space. Several other kinds of bacteria also undertake some action only when their number reaches a critical level. For many of these, the responsible signals belong to a group of chemicals called homoserine lactones that are chemically similar to *Vibrio fischeri*'s pheromone, N-(3-oxohexanoyl)homoserine lactone. These signals are often called quorumsensing pheromones, because the bacteria can take action only when a quorum is present.

Quorum sensing also triggers events more elaborate than the production of light. One of the most remarkable of these involves rod-shaped microbes known as myxobacteria, such as *Myxococcus xanthus*, that flourish in cultivated soil all over the world. These bacteria live individually in the soil as long as food is in good supply. If water or nutrients begin to fail, about one hundred thousand cells come together, progressing through the soil to a gathering point. Here the cells develop an elaborate structure known as a fruiting body, within which they undergo a remarkable transformation. Over the next twenty-four hours, they turn themselves into spores. Unlike the free-living cells, these spores are seedlike thickwalled structures that are resistant to heat, starvation, and lack of water. Although the individual cells are microscopic, the mature fruiting body they create is just large enough to be seen with the naked eye as a colored speck. Not all of the aggregated cells become spores, as many of them are sacrificed in constructing the fruiting body.

In taking this drastic action, the bacterial cells collectively desert a locale where nutrients have become scarce. Now a packet of spores, they await transportation to a new home. The wind, an animal, or perhaps flowing water will pick up the fruiting body and deposit it elsewhere. The spores of course do not guide their journey, but if by chance they land in an appropriate environment, they then revert to their free-living form. If nourishment is plentiful, they may establish a flourishing new colony of bacteria.

Transforming free-living myxobacteria into a fruiting body requires at least four different chemical signals. Two of these are well enough understood that we can describe them briefly. The first is a quorum-sensing pheromone that promotes the initial aggregation of individual cells. When a cell no longer finds adequate nutrients, it secretes a mixture containing several common amino acids. This mixture spreads through the soil, broadcasting its message in all directions: "I am here, and I am starving." As long as only a few cells are sending this message, the pheromone's concentration in the soil remains low. If many cells begin to signal a lack of food, it naturally rises. The colony members sense this concentration in their surroundings. While it is low, they take no action: Most cells still have adequate nourishment. Aggregation commences only when the pheromone level indicates the number of starving cells is sufficient to assemble a complete fruiting body. The signal now declares not only that nourishment is scarce but also that the number of protesting cells is great enough to take meaningful action. By aggregating only when assured they can form a fruiting body, the bacteria increase the probability of successful relocation.

Once the cells come together, they begin to produce a second pheromone, which activates spore formation. Unlike the first signal, this pheromone is a small protein that remains attached to each cell's surface. Since it does not spread through the soil, cells must be in direct contact to distribute its message. During aggregation, each rod-shaped cell moves about, adjusting its position so the cells fit together snugly, end to end and side by side. As the cells become aligned, the signal passes from one cell to the next, ultimately reaching the entire mass. Only then do they begin the transformation into spores.

The requirement that cells be in intimate contact to transmit the second pheromone ensures that spore formation commences only after they have arranged themselves into a compact mass. Closely packed cells generate a fruiting body filled with closely packed spores. The arrangement is critical because the likelihood of successful relocation depends on the number of spores the fruiting body contains and how efficiently they are packaged. Although the two remaining chemical signals are not yet well understood, we already know enough to appreciate these microbes' impressive communal effort to perpetuate their species.

The lifestyles of many bacteria depend on maintaining a close affinity with more complex organisms. Some of these microbes are serious pathogens, whereas others are harmless creatures that pursue quiet inoffensive lives unnoticed by their hosts. Still others provide significant advantages to their hosts, much as do nematode-associated bacteria, and in at least certain cases these microbes and their hosts exchange chemical messages. One of the most important and best-studied groups of such beneficial bacteria are those that can fix nitrogen. That is, they possess the rare capability of taking up atmospheric nitrogen gas (a molecule consisting of two nitrogen atoms bonded together) and converting it into ammonia (a nitrogen atom bonded to three hydrogen atoms). The significance of this process derives finally from a universal biological necessity: Plant or animal, one-celled or complex, every living creature must have available useable forms of nitrogen for synthesizing proteins, nucleic acids, and other chemicals essential for life as we know it.

For animals, the ultimate sources of useable nitrogen and all other nutrients are plants and algae. Animals that consume these organisms ingest nitrogenous compounds as they feed. Those that eat other animals participate in food chains that lead eventually back to these organisms. As for plants, most of them assimilate nitrogen-containing minerals (and fertilizers, if they are cultivated species) from their environment. Because the earth's atmosphere is largely (78 percent) nitrogen gas, you might suppose that plants long ago developed biochemical machinery to convert this immense resource into useful nitrogenous compounds. They have not done so, however, probably because making direct use of atmospheric nitrogen comes at a high price. The nitrogen molecule is particularly stable, and converting it to a biologically useful form is energetically very expensive. Even epiphytes, or air plants, make no use of nitrogen gas, although they live out of the soil anchored to other plants, often with their roots permanently exposed to the air. Only nitrogen-fixing bacteria have mastered the biochemical feat of converting nitrogen from the air into a form useable by other species, which ammonia is.

Though plants cannot take direct advantage of plentiful atmospheric nitrogen, some have done the next best thing. Like traders acquiring from abroad goods they cannot produce at home, these plants have established beneficial relations with certain nitrogen-fixing bacteria. These bacteria are called rhizobia and belong to the genus *Rhizobium* or one of its close relatives. Nearly all of the host plants are legumes, such as peas, beans, and clover. (We shall collectively call these hosts legumes, even though a few other species are included.) This rhizobium-legume association allows the plants to appropriate for their own biosynthetic needs ammonia synthesized by the microbes. By guaranteeing the legumes a continuing supply of useable nitrogen, the bacteria free them from dependence on soil nitrogen for sustenance. In return, the legumes supply nutrients and a stable environment to keep the bacteria well nourished. The association also figures in the complex global cycle by which nitrogen continually passes back and forth among minerals, living systems, and the atmosphere. Other paths also remove significant amounts of nitrogen from the atmosphere, such as the large-scale manufacture of nitrogen-containing fertilizers, but the association between plants and bacteria accounts for perhaps a quarter of the nitrogen fixed worldwide.

In establishing their relationship, bacteria and plants engage in an extensive chemical dialog. Rhizobia live not only with plants but are free-living soil bacteria as well. Free in the soil, rhizobia do not fix nitrogen, although there are other soil bacteria that do. When free rhizobia find themselves near a legume root, they may detect amino acids, sugars, and other attractants released from the plant's tiny root hairs. Drawn to this nourishment, the rhizobia move toward the root. Once the bacteria are nearer, they pick up another root hair signal that has a more profound effect. The exact chemical nature of this second signal varies among legumes, but in all known cases, the signals are related to chemicals known as flavonoids, which include many common plant pigments. A plant's particular flavonoid signal is one factor in determining its associated rhizobia, although there is no one-to-one correlation of species in the interaction between plant and rhizobium. Some organisms, both plants and rhizobia, associate with several different partners, and others with only one.

When the flavonoid signal reaches the rhizobia, it induces them to produce and secrete a compound composed of sugarlike units that is called Nod (for nodulation or nodule-forming) factor. Nod factor spreads through the soil and is soon detected by the root that sent out the flavonoid. Here its message to the plant is to begin building a root nodule, which in time the rhizobia will inhabit. Nod factor induces the cell division in the root necessary to form a nodule and at the same time causes the root hairs to grow, branch, and become somewhat deformed. As this is happening, the rhizobia are moving closer, soon coming into direct contact with the root hairs. Carbohydrates borne on the bacterial surface now signal the root hairs to develop tiny tubules called infection threads. These carbohydrates then enable the rhizobia to pass into these infection threads. Once the rhizobia enter the threads, they are inside the plant. While plant cells are proliferating in the root to create a nodule, rhizobia begin proliferating within the infection threads.

Subsequent events are visible under a microscope, but information about the signals involved is vague. Very likely there is an extended exchange of chemical messages both between bacterial and plant cells and also among the bacteria themselves, because the ensuing events demand close coordination. The rhizobia, increased in number through division, move down the tunnel-like infection thread. On reaching its end, they induce a weakening in plant cell walls, probably through the agency of another chemical secretion, and make their way into cells of the developing nodule. Once inside, the rhizobia undergo a fundamental transformation. They increase in size and differentiate into what are called bacteroids, cells that no longer undergo cell division but begin to fix nitrogen. The plant cells maintain the bacteroids, providing them with nutrients and nitrogen to fix, and carefully controlling the local acidity and other conditions—the price the plant must pay to benefit from nitrogen fixation. As the bacteroids generate ammonia, the plant cells assimilate it for their own use.

In addition to uncertainties about chemical signals in these latter steps, many other questions about this extraordinary relationship are still unanswered, including some quite basic ones. For instance, details of the exportation of ammonia from bacteroid to plant cell remain unclear, although this is one of the relationship's principal events. On another level, it is debatable how the interaction is beneficial to the rhizobia. There is no doubt that it serves legumes admirably, for it allows them to extend their range into nitrogen-poor soils where other plants cannot survive. The advantages to the bacteria, however, are less obvious. Their differentiation into bacteroids seems to be a terminal event. Thereafter, these cells remain within the plant fixing nitrogen, but they no longer proliferate. Their association with the legume gives them no obvious reproductive advantage. Yet, for such a complex interaction to evolve and persist, both partners must find some advantage in it. The interaction expends energy that must afford some gain; otherwise, it is a losing proposition.

At present there is only speculation about benefits to the bacteria. One proposal is that a small portion of rhizobia are not transformed into bacteroids, and that living within the nodule, they enjoy a reproductive advantage over rhizobia remaining free in the soil. Another possibility is that rhizobia within a plant root export nutrients to their relatives in the free colony outside and so better their chances of survival. There is no firm evidence to support either of these suggestions or any others. This is an area where puzzles still abound.

Nitrogen-fixing bacteria are important to humans as well as to their legume partners. Farmers raise such leguminous crops as soybeans and alfalfa in soil that would not support wheat or corn. Also, legumes are the key to crop-rotation schemes that prevent repeated plantings from exhausting cropland. Wheat, corn, and other grains depend on nitrogen they take from the soil. Planting these crops on the same land year after year soon depletes the soil of nitrogen, leaving it less productive. However, if grain is alternated every few years with legumes, the soil's nitrogen can be restored. Since leguminous plants take their nitrogen not from the soil but indirectly from the atmosphere, plowing these plants under at the end of the growing season enriches the soil with their store of nitrogenous compounds. This usable nitrogen then becomes available to support another round of grain crops.

Crop rotation has been a widespread practice in Western agriculture for more than two hundred years. In the future, nitrogen fixation may bring another sort of agricultural improvement. Now that the techniques of genetic engineering permit biologists to transfer functioning genes from one organism to another, it may be possible to introduce rhizobial associations into plants where they do not exist naturally. Plant scientists at the International Rice Research Institute (IRRI) in Manila have undertaken research with rice to do just this.

This research program is one of the Institute's several long-range endeavors focused on augmenting the world's rice supply. IRRI is a nonprofit agricultural research center "dedicated to helping farmers in developing countries produce more food on limited land using less water, less labor, and fewer chemical inputs, without harming the environment." Since 90 percent of the global rice crop is grown and consumed in Asia, the significance of these efforts may be difficult for us in the West to appreciate fully. Perhaps it helps to note that a third of the people on earth depend on rice for more than half their daily food and that a large percentage of these people live in subsistence-level poverty.

If rice were able to host rhizobia and assimilate the ammonia they produce, its range could be extended to nitrogen-poor land that cannot now support it. It is much too early to know whether the necessary genes can be introduced successfully into rice and so enable it to interact with rhizobia as legumes do. The research is an ambitious undertaking, but for IRRI developing a new type of rice is a goal well worth the effort. The current estimate is that rice production must rise by almost 70 percent over the next thirty years to meet projected growth in demand. Maybe bacteria can help.

In closing, we should note that bacteria are already helping in a broader way, not in farmers' fields but in research scientists' laboratories. As it has become clear that these simple creatures employ special chemicals much as multicellular organisms do, bacteria have become attractive laboratory subjects for fundamental studies. Although each of these organisms consists of only a single cell, they steal chemical compounds, share them with others, and communicate using them, much as more complex creatures do. Probing the details of such processes is often more easily undertaken in simply organized creatures rather than more complex ones. On a molecular level, the remarkable chemical abilities of bacteria have begun to teach us a great deal about how all organisms exploit special chemicals.

CHAPTER

Delving into Nature's Chemicals

8

For thousands of years, humans have devised practical uses for chemicals found in nature. Egyptian priests depended on natural oils and resins as antibiotics in mummifying human and animal remains. South American Indians prepared plant extracts to tip their arrows with fast-acting poison. The ancient Chinese discovered the secret of silk, magically converting an insect's cocoon into an elegant textile.

Each of these discoveries resulted from turning special chemicals to human needs. Such developments are one of the most significant connections between these natural compounds and our own lives. After exploring special chemicals in their natural setting, we should now examine their practical effects in human life. Some of these discoveries have been with us for thousands of years, and new ones are being made every day. By looking at both ancient and modern discoveries, we can appreciate the amazing range of these developments and what they have contributed to our welfare.

Primitive chemical discoveries included both effective crude mixtures of chemicals and also relatively pure compounds, as well as intact organisms utilized for their unique chemicals. Beyond using chemicals directly in this fashion, modern humans have developed other ways of exploiting them. Genetic engineering has converted ordinary laboratory bacteria into chemical factories, creating microbes with altered genes that tirelessly churn out other organisms' chemical creations. In this way, insulin and several other pharmaceuticals are manufactured in large quantity for clinical use. In another direction, scientists investigate how organisms synthesize materials necessary to their lifestyles. The understanding gained may point the way to developing novel products in the laboratory. Today, these and other means of benefiting from nature's chemicals have evolved into a huge commercial enterprise with annual sales of more than $10 billion in the United States alone. This research-backed enterprise brings humans and natural compounds together and includes activities from several closely related areas of technology. We can ignore the formal distinctions among them and for convenience call the entire endeavor biotechnology. This activity is a significant segment of our story, and we can explore it through examples, old and new.

One of the most successful biotechnological industries of the ancient world came from the Phoenicians (the Biblical Canaanites). In one of history's longest-lasting business ventures, they prepared and sold at great profit the dye we know as Tyrian or royal purple. Beginning about 1600 BC, along what is now the Lebanese coast the Phoenicians established their first dye works. Tyrian purple imparted a lovely reddish violet hue and was the fastest dye for wool known to the ancient world. It was expensive and so became a much-sought-after symbol of elegance and wealth. As the market for Tyrian purple grew, the Phoenicians expanded manufacturing to other suitable sites. They sold their purple dye throughout the Mediterranean wherever they could reach by ship and sent it overland all through the Middle East.

They found eager customers in Egyptians and Etruscans, and later among the Greeks and Romans. Jewish traditions accorded Tyrian purple an important place, and there are several Biblical references to it. The demand for the dye exceeded the supply for centuries, guaranteeing that the price was always high. At times through its long history, Tyrian purple sold for ten to twenty times its weight in gold.

The Phoenicians protected their monopoly for centuries by successfully concealing the source of Tyrian purple. Eventually, the secret came out. In about A.D. 60, the first full-length account of preparing the dye appeared in Pliny the Elder's *Natural History*. Pliny reported that Tyrian purple came from sea snails, and he described the Phoenicians' manufacturing process to his readers. As he described in somewhat different terms, several widespread marine snails (*Murex trunculus* and related species) produce a secretion containing the precursor chemicals of the dye. On exposure to light and oxygen, this mixture undergoes a series of slow chemical transformations that lead to Tyrian purple.

The dye's origin sounds obscure, but discovering it was probably simple enough. The snails were a common food item. In preparing them for the table, people must have frequently come upon the snails' so-called hypobranchial gland and its dye-forming secretion. Often they must have inadvertently stained their hands while holding the snails and thus learned of the dye by chance. The Phoenicians were not alone in making this discovery. The snails are common in coastal areas around the world, and both the ancient Chinese and Peruvians independently discovered and used the purple dye. In modern times, Lebanese children playing on the beach not far from early Phoenician dye works are said to dye rags by dipping them into crushed snails.

As Pliny reports, the Phoenicians caught snails for their dye works in baited wicker baskets. In manufacturing the dye, they removed the hypobranchial glands and heated the glandular secretion in large vats of salt water for ten days. Treating about ten thousand mollusks in this fashion yielded only one gram of Tyrian purple—hence the dye's rarity and high cost. The Phoenicians set up factories all around the Mediterranean wherever they could find large beds of snails. Mounds of uncounted millions of crushed shells still remain along the seacoast, attesting to their lucrative enterprise.

The expense of Tyrian purple ensured that only the wealthy could afford it. In time, it became the especial prerogative of high state officials and ecclesiastics. With the advent of the Roman Empire, access to Tyrian purple grew even more restricted. Nero decreed in the first century that only the emperor might wear Tyrian purple garments. During his reign, the penalty for violating this edict was death, although the law seems to have been disregarded increasingly under later emperors. In the fourth century, Rome nationalized the industry and required that all Tyrian purple be manufactured in imperial dye works. The Tyrian purple industry prospered until the decline of Rome in the fifth century but finally disappeared only with the Ottoman invasion in the fifteenth century.

Long before the Phoenicians, humans had adopted structural materials from the creatures around them. From such sources as hides and plant fibers, they fashioned goods to meet new needs as they arose. Somewhat analogously, there is continual pressure in our complex society for new materials having unique properties. Surgeons demand strong, light substances compatible with the human body for replacing knee joints and repairing broken hips. Dentists seek ceramic-polymer composites for tooth-colored fillings and crowns that will last a lifetime. The U.S. Army sponsors research into rugged lightweight materials such as spider silk for lining bulletproof vests. Rocket engines depend upon structural elements that are stable at extremely high temperatures, whereas special lubricants for spacecraft should function faultlessly in extreme cold. Efforts to satisfy these endless needs have spawned several lines of research, including a new branch of biotechnology. Materials scientists have joined with chemists and biologists to investigate structural materials from nature. Several organisms make glue better than ours, and spiders spin silk stronger than steel. Perhaps we can profit from such biological creations, but first we must understand the chemical and physical bases of their exceptional properties.

Structural problems in the living world are of course enormously varied, but their solutions need not be complex. Seed corn, for example, owes its remarkable long-term viability to a surprisingly simple sealant. As farmers know, after fifty years of dry storage kernels of corn still sprout when planted in moist soil. What protects the dormant seeds' sensitive enzymes and store of nutrients from degradation over decades of exposure to the air? Despite years of agricultural research, this elementary question had no answer until the late 1980s. We now know that corn kernels have an impermeable outer coating composed of ordinary table sugar (sucrose) and a closely related compound (raffinose). As long as the corn is dry, this coating affords a hard glassy shield that seals the kernels from the air. When the corn is moistened, the sugars quickly dissolve, allowing the kernels to absorb water and sprout.

This simple coating has found its way into a novel drug-delivery system. Drug developers have adapted the idea to create an inhalable form of insulin, offering a startling new way of delivering this medication for diabetes. They prepared a mixture of insulin (a small protein) and sugars in solution and then dispersed the liquid into a mist of fine particles by spraying it through a nozzle. The airborne particles quickly dried to give a protected insulin powder, where each particle consisted of a small speck of insulin sealed within a sugar coating. The coating protects the particles just as it does kernels of corn, allowing insulin to be stored without deterioration. When a diabetic inhales the powder, it passes into his lungs. The sugars dissolve in the moist air sacs, and the specks of pure insulin that remain are quickly absorbed into the blood stream. This could be a much more agreeable way of taking insulin than the customary injection. Clinical tests of inhalable insulin are already under way.

Of course, many natural materials are more difficult to reproduce in the laboratory than the sugar coating on corn kernels. One that still eludes materials scientists comes from bacteria that possess a magnetic sense. Just as many living creatures have a visual sense that responds to light, many also have a sense that responds to magnetic fields. In their natural setting, these creatures sense the earth's magnetic field, registering its intensity or orienting themselves with respect to it. Motion in response to the field is called magnetotaxis, and the best studied magnetotactic organisms are several kinds of bacteria found in freshwater or marine sediments. One of the freshwater species goes by the expressive name *Aquaspirillum magnetotacticum*.

Magnetic microbes were unknown until 1975, when an observant graduate student discovered them under his microscope. While he was studying bacteria from the bottom of a muddy marsh, he noticed they consistently moved to one side of the microscope slide. On investigation, he found that no matter how he oriented the slide, where he placed the microscope, or whether it was in the light or dark, the bacteria responded in the same way. Whatever the conditions, the bacteria always swam north. The extensive investigations prompted by this intriguing observation revealed that each microbe is a miniature compass needle. Each one contains a chain of crystals of the common magnetic mineral known as lodestone or magnetite (a form of an iron oxide), and thus each tiny crystal is a permanent magnet. Because the crystals making up a chain are magnetically aligned, the aggregate behaves like a compass needle. Extending along the length of the cell, it serves to point the bacterium northward and determine the direction it swims.

What do bacteria from a muddy marsh bottom have to gain by invariably swimming north? The microbes actually have no interest in north or south, but their magnetic sense also allows them to distinguish up from down, which for them is a meaningful distinction. If you are in the Northern Hemisphere, you may have noticed that a compass needle not only points north but also downward. This is because the earth's magnetic field curves up and outward from the south magnetic pole and down and inward into the earth at the north magnetic pole. In the northeastern United States, where the original marsh-mud bacteria were collected, the field's down-

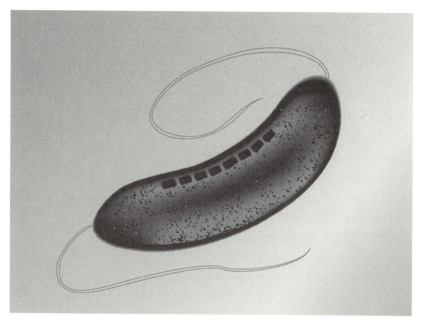

ILLUSTRATION 12 This bacterium
carries a line of magnetite-containing
particles that functions as a compass owing
to the precisely controlled size and
orientation of the particles.

ward angle toward the north magnetic pole is about 70 degrees.
Just as a compass needle points downward, the magnetic particles
orient the bacteria downward as well as to the north. This is important, because it keeps the bacteria near the bottom sediments,
where there is relatively little oxygen. Like many other microbes,
these bacteria thrive only in a low-oxygen environment. If they
swam upward, the higher level of oxygen at the water's surface
would poison them. It is vital for them to know which way is up.

Owing to the shape of the earth's magnetic field, a compass
needle points downward only in the Northern Hemisphere. In the
Southern Hemisphere, the field's curvature causes the needle to
point north and upward. Here, the south end of the needle points
downward. As a result, the magnetic sense of New Zealand bacteria works in the opposite direction. They too should remain near

the bottom sediments, and so these antipodal microbes swim south instead of north.

Materials scientists find the small crystals at the heart of this magnetic sense fascinating. They can grow magnetite crystals in the laboratory but find it exceptionally difficult to grow ones of the size found in bacteria. The bacterial crystals' long dimension is 40–120 nanometers. A nanometer is one one-millionth of a millimeter, or roughly one-thousandth the diameter of a fine human hair. Size is critical, because it dictates the crystals' magnetic properties in an aggregate. If the crystals are smaller than 40 nanometers or much larger than 100 nanometers, their aggregate does not function as a magnet. Only intermediate-sized crystals—the size the bacteria contain—can aggregate to behave like a compass needle.

Bacteria have the right-sized crystals because the iron dissolved in their cells probably crystallizes within tiny compartments. The currently favored explanation is that the microbes fashion large molecules into a matrix with uniform elements that have the dimensions of the desired crystals. As magnetite (iron oxide) crystallizes from solution, the growing crystals are confined within these elements and can attain only the appropriate size. This process of matrix-directed crystallization is something we cannot carry out nearly so well as these single-celled organisms. Matrices also govern the biosynthesis of structural materials as disparate as bone and mother-of-pearl. The synthetic processes leading to these substances are complex in detail and not yet particularly well understood. If materials scientists could emulate the bacterial crystallization of magnetite, industrial applications of this new biotechnology should soon follow. Microscopic magnetic particles ought to find ready application in preparing improved magnetic coatings for such goods as motors, loudspeakers, and magnetic tape.

Many other creatures also possess a magnetic sense. Certain insects, turtles, and fishes, as well as some birds and mammals, respond to the earth's field. Several of these organisms contain localized magnetite crystals as well, but in only one vertebrate is there firm evidence linking the crystals to a magnetic sense. Like many vertebrates that detect the earth's magnetic field, rainbow trout (*Oncorhynchus mykiss*) use this ability as an aid to navigation. The fish carry their magnetite in a small region of their nose, from which sensory nerve fibers pass straight to the brain. This same nasal region has also been linked experimentally to the fish's magnetic response.

A promising material of a quite different sort comes from the edible blue mussels (*Mytilus edulis*) that are common along seashores around the world. To avoid being tossed about in the water, blue mussels anchor themselves to a rock or other holdfast with fibers known as byssal threads. These coarse, dark-colored strands, popularly called a mussel's beard, are strong modified tendons. Like other tendons, byssal threads are composed largely of a widespread structural protein called collagen, but they are significantly sturdier than human tendons and much more elastic. These differences are important. Mussels live along coasts in the intertidal zone, where waves batter them endlessly. To withstand the constant assault, their attachment must be both secure and flexible or they risk being torn loose by the surf and swept away.

Byssal threads combine strength and flexibility in a novel way. A thread is elastic near the mussel's foot but it is stiff at its other end where it attaches to the anchoring holdfast. This permits it to act as a shock absorber close to the mussel, and at the same time act as a tough tether at the holdfast. In between, the thread's properties vary gradually, as it becomes progressively firmer and less elastic from the mussel to the holdfast.

These characteristics are a direct consequence of the collagen's chemical structure. Close to the animal, these molecules consist of normal collagenlike structural regions flanked by elastic regions, an arrangement that allows the thread to stretch as much as 160 percent. At the holdfast end, the elastic regions are absent and are replaced by fiberlike regions. Here all the material is nonelastic and more like typical collagen, which is only about 10 percent stretchable. Intermediate regions are as yet unexamined, but presumably the chemical structure of the threads mirrors their properties, changing along the length of the thread gradually from one end to the other, with fiberlike regions increasingly replacing elastic ones.

The chemical picture of this modified collagen is not yet complete, but its mixed structural regions already hint at similar new synthetic materials. For the NIH, which has supported this biotechnological research, the hope is that this unique material, with its variable elasticity, will lead to more comfortable and pliable artificial skin. Ultimately, it could also improve the quality of such goods as footwear and radial tires, where softness and toughness are equally desirable.

Centuries before anyone was interested in the chemical properties of mussel collagen, another creature's byssal threads attracted praise and admiration. In Roman times, a large mollusk known as the noble pen shell (*Pinna nobilis*) was quite common in the Mediterranean. Occasionally reaching a length greater than a meter (more than three feet), it is one of the world's largest mollusks. Noble pen shells live partially buried in the sea bottom, attaching themselves to a holdfast through byssal threads, just as blue mussels do. However, the threads themselves are quite different in the two species. Unlike blue mussels' rough fibers, the byssal threads of noble pen shells are long and slender. They appear delicate but are much stronger than they look.

Whereas modern scientists study byssal threads as a chemical model for new materials, ancient Romans valued them as an article of commerce. They collected pen shells specifically for their particularly fine byssal threads, and skilled weavers worked these filaments into a wonderfully sheer fabric. This was a fabric so light, so exquisite, that it was known fancifully as linen mist. It must have been lovely, for there were those who insisted that the famed Golden Fleece of mythology was not wool at all but actually the stuff of linen mist. Needless to say, wiser folk ridiculed this idea, knowing that the Golden Fleece had been taken from the Golden Ram and had nothing to do with shellfish. Unfortunately, linen mist has disappeared from the world, for it is no longer woven in

Rome or elsewhere. Noble pen shells have become so scarce that in Italy and other Mediterranean countries they have the status of a protected species. Possession of a single freshly taken specimen can lead to a stiff fine.

The Romans made direct use of byssal threads from noble pen shells. This ancient practice of using natural chemicals directly without modification remains the most widespread form of biotechnology. A recent innovation in this area is the expanding market for enzymes that function under extreme conditions. To appreciate the industrial value of these compounds, we must first briefly discuss enzymes and then consider some exotic organisms.

An enzyme is a protein that speeds up a biochemical reaction without itself experiencing any overall change. In chemical language, such a compound is called a catalyst and is said to catalyze a reaction. Chemists employ a variety of compounds as laboratory catalysts, and many industrial chemical processes would be impracticably slow without catalysis. An automobile's catalytic converter makes use of a metal catalyst to accelerate conversion of toxic carbon monoxide in the exhaust to carbon dioxide. Similarly, our bodies' biochemical machinery effects thousands of different reactions that would not proceed without enzymatic catalysis. Some enzymes are exquisitely specific, catalyzing only one particular reaction of a single compound. Many others have much less exacting requirements and consequently exhibit broader effects. Specific or nonspecific, enzymes can make reactions go many millions of times faster than they would without catalysis.

In recent years, the industrial market for these biological catalysts has expanded briskly. In comparison with traditional chemical, nonbiological catalysts, enzymes frequently offer more specific catalysis and have a less deleterious impact on the environment. Available either directly from living creatures or through genetic engineering, industrial enzymes now comprise a worldwide market estimated at $2.5 billion. In such diverse operations as drug production, food processing, and the manufacture of detergents, they have become indispensable.

Ordinary enzymes are big business, but they would be employed even more widely if they were not so fragile. An enzyme's catalytic activity derives from its exact molecular shape, which is easily disordered. Like other proteins, an enzyme molecule is a long chain of amino acids linked one after another like the cars of a train. Unlike a train, however, a protein molecule is not straight. When it is synthesized in a cell, a long amino-acid chain folds up in a characteristic fashion to give the molecule its final three-dimensional form. In this form the chain will doubtless have many twists and turns, perhaps including occasional regions where portions of the chain double back and forth on themselves and, lying side by side, form flat sheets. Adjacent portions of the same chain may spiral into long corkscrews or undergo sharp turns. Whatever the details, in a natural cellular environment, numerous weak chemical interactions occur between parts of the chain that are folded into close contact with one another. Such interactions serve to stabilize, or lock in, the three-dimensional form.

These stabilizing interactions are destroyed if an enzyme is exposed to conditions much harsher than its natural ones. Thrown into industrial processes where temperatures are too high or a mixture is too acidic or too alkaline, enzymes will unfold and immediately lose the shapes necessary for their catalytic activity; the chain of amino acids has not necessarily broken, but the stabilizing interactions are lost and the unfolding molecule becomes a messy conglomeration of disordered, catalytically useless arrangements. Once this happens, many enzymes cannot be coaxed to refold into their catalytically active form. This sensitivity limits industrial use of enzymes to processes carried out under suitably mild conditions.

The irreversible loss of a protein's native molecular shape is familiar to anyone who has boiled an egg. The white of an egg is largely a single protein called albumin. In a fresh egg, each albumin molecule is folded in a particular way that is its natural shape. This arrangement of each protein chain is stable at room temperature, but heat disrupts the interactions holding it together. At the temperature of boiling water the albumin unfolds, becoming a jumble of molecular chains that clump together and solidify. No one has contrived a way of restoring the white of a boiled egg to its original folded form.

Some years ago, the fragility and limitations of conventional enzymes led to a concept new to biotechnology: the industrial employment of enzymes taken from organisms that live in more demanding environments than do the majority of creatures. How fragile could an enzyme be if it came from an organism native to an unusually hot environment? Such an organism's components would necessarily function normally in its habitat, even if most other creatures could not tolerate the high temperature. This idea was enticing because there are microorganisms (and at least a few larger species) that inhabit such harsh environments. These creatures remained undiscovered until about forty years ago, but the search for new ones living in areas of high natural temperatures has intensified in recent years, spurred on by the possibility of industrial applications. Biologists now appreciate that environments once presumed to be sterile are in fact teeming with life. These hardy microbes are customarily called extremophiles, although this name and the notion that their habitats are extreme only reflect our own parochial point of view. The organisms themselves evidently find their living conditions quite normal.

Because extremophiles are often difficult to obtain, their enzymes are prepared in industrial quantity through genetic engineering techniques. A small amount of an extremophile can furnish a sample of the gene associated with the biosynthesis of a desired enzyme. Molecular biologists transfer this gene from its source into laboratory bacteria which then follow its instructions and synthesize the extremophile's enzyme in commercial quantities. Typically, there is no further need to return to the extremophile for more of the native enzyme.

One of the most important of these extremophiles is a bacterium (*Thermus acquaticus*) discovered in 1965 in a Yellowstone National Park hot spring where the temperature is a constant 73 °C (Centigrade or Celsius) or 163 °F (Fahrenheit). About twenty years later, this microbe contributed an enzyme to a new procedure that ultimately grew from a laboratory curiosity into a practical process. It has had far-reaching effects that are now well known around the globe. This process is the so-called polymerase chain reaction (PCR), which amplifies a few strands of DNA into many millions. It allows the tiny bit of DNA in, say, a drop of blood or saliva to be augmented enormously and so furnishes sufficient material for a variety of purposes. A procedure that amplifies DNA is essential in genetic engineering and for other laboratory applications, but the more newsworthy use of PCR is the forensic examination of DNA samples. Properly executed, this test can determine to high probability whether two samples of DNA are derived from the same individual. The test achieved international renown through the DNA identifications introduced and then challenged during the O. J. Simpson murder trial. Only a few years later, PCR returned to the world stage for a titillating second appearance, when it pointed unequivocally to the source of the infamous stain on Monica S. Lewinsky's blue dress. Through many less celebrated applications, comparisons of DNA samples have increasingly become admissible as evidence in courtrooms across the land.

Originally, PCR was a tedious experiment. The key component is a ubiquitous enzyme called DNA-polymerase that makes it possible to copy pieces of DNA over and over. The procedure requires alternate heating and cooling of a mixture of enzyme, DNA, and the four nucleotide building blocks that are used to synthesize more DNA. The DNA-polymerase initially employed in PCR came from an ordinary bacterium. Because the enzyme was not heat-resistant, however, it lost its activity each time the mixture passed through the procedure's hot cycle. A technician then had to add fresh enzyme before the next cycle could begin. PCR experiments proceeded slowly and demanded constant attention.

To overcome this shortcoming, biologists turned to an enzyme that could survive the PCR hot cycle. They replaced the original heat-sensitive enzyme with DNA-polymerase from *Thermus acquaticus*, the Yellowstone extremophile. The new enzyme is unscathed by the high-temperature phase of the PCR procedure and continues functioning throughout the experiment. With this modification, PCR became much more convenient. The entire procedure could be automated, and it rapidly became an essential laboratory tool for molecular biologists. More recently, an even more heat-stable DNA-polymerase has become popular for PCR. This enzyme is from a marine microbe named *Pyrococcus furiosus* and works best at 100 °C, the temperature of boiling water.

As the search has expanded in the last few years, biologists have discovered many microorganisms flourishing in high-temperature environments. One rich source has been hydrothermal vents, which are rock chimneys rising from the ocean floor. Mineral-rich liquids welling up through these vents from deep within the earth erupt into the ocean at temperatures as high as 400 °C (752 °F). These fluids pass directly from the vents into water that is just above freezing, mixing minerals from far beneath the ocean with chemicals already present in the water. This mingling generates environments at several different temperatures, each characterized by an unprecedented mixture of chemical compounds. An incredibly rugged microbe from one of these habitats is *Pyrolobus fumarii*. Accustomed to life in a hydrothermal vent wall, it does best at about 105 °C and continues to reproduce even at 113 °C. Below about 90 °C (194 °F), it fails to grow because it is too cold. The upper limit of temperature for these hyperextremophiles will probably be about 150 °C. Higher temperatures interfere with vital functions, as sensitive chemical bonds in DNA itself and in other essential molecules begin to rupture irreversibly.

Enzymes from these organisms find numerous applications beyond PCR. Certain ones can modify plant fibers or break down proteins or fats. Heat-stable enzymes with these digestive properties are very attractive to the food processing industry. Properly controlled, their action can make prepared foods more palatable, and they can be employed where ordinary enzymes fail. In food processing, all operations must be carried out under sterile conditions. Frequently, the easiest way to maintain sterility is to keep the temperature above 60° C at all times. This is hot enough to disable conventional enzymes, but heat-stable ones of course do not have this limitation. Additional high-temperature applications exist in drug development and for recovering residual oil from oil wells.

Extremophiles also flourish in frigid environments where most organisms would freeze to death. These cold-adapted organisms inhabit vast portions of the oceans where the temperature averages 1–3 °C (34–38 °F), as well as polar regions that can be even colder. Nonetheless, extremophile bacteria, algae, and protozoa thrive in these environments, just as their less robust relatives do in temperate waters. Mirroring the cold sensitivity of heat-resistant organisms, these cold-resistant species can tolerate very little heat. One antarctic bacterium (*Polaromonas vacuolata*) prospers at 4 °C (39 °F) but cannot reproduce in surroundings above 12 °C (54 °F). Typical bacteria are comfortable at temperatures in the range 20–40 °C.

A number of applications look ready-made for cold-adapted enzymes. Those that degrade protein and fat appeal to the detergent industry to enhance cold-water washing products. Fragrance and perfume producers constantly want new catalysts for low-temperature manufacturing processes. They prefer to operate in the cold, because precious odor ingredients evaporate quickly at higher temperatures. The food processing industry is interested as well, as its work frequently must be carried out at refrigerator temperature to retard spoilage.

Beyond their practical value, extremophile enzymes present scientists with a fundamental puzzle. Like all molecular characteristics, their exceptional stability must originate in their chemical structures. However, it is not yet certain what structural features determine these properties. What is known is that in their active folded form, cold-resistant enzymes appear to have relatively fewer structure-stabilizing interactions between different parts of the amino acid chain. As a result, they remain more flexible at a lower temperature than ordinary enzymes but unfold and lose their activity more quickly as the temperature is raised. Conversely, heat-resistant enzymes seem to have a larger number of stabilizing interactions than their conventional counterparts and thus retain their folded structures at higher temperatures. At normal temperatures, however, the increased interactions render them relatively inflexible, and they no longer function efficiently.

Further insight will come only with structural studies on more of these odd catalysts. As understanding improves, protein scientists will be eager to apply this new knowledge to introduce structural changes in ordinary enzymes that would enhance temperature stability and catalytic efficiency. For the present, the microbes do it better.

Stocking the Medicine Chest

Prehistoric peoples exploited nature's special chemicals in impressive ways, and perhaps the oldest of these is to apply these chemicals as medicines. As we shall see, in our modern world this activity has grown into a multibillion-dollar pharmaceutical industry that is responsible for saving thousands of lives each year. This notion of deriving medicinals from nature runs throughout human history and may even predate any early human activity. Consider this brief report: A chimpanzee (*Pan troglodytes schweinfurthii*) that seems to have a stomachache chews avidly on bitter stems from a bushy plant she normally avoids. Indigenous people have long prescribed the same plant for treating intestinal upsets. Within a day or two, the chimpanzee recovers and rejoins her family.

Similar apparently self-medicating behavior involves the scarlet macaw (*Ara macao*), a large, brightly colored parrot species found in the rainforest of southeastern Peru. These birds are fond of the poisonous unripe fruit of the sandbox tree (*Hura crepitans*), tearing it open with their powerful beaks to feast on its flesh and seeds. They survive eating the poisonous fruit only because they also eat a clay they find on high river banks that neutralizes the fruit's toxin. The macaws eat this clay regularly and feed it to their chicks, who clearly relish the treat and clamor for more.

Intriguing observations of this sort have been accumulating for more than twenty years. How are we to decide whether the chimpanzee was literally medicating itself? If she was, how did chimpanzees commence using plant medications? Did scarlet macaws learn to detoxify the poisonous fruit in their diet by eating clay and then safely consume even more poison? Or did the clay already in their diet permit them to begin feeding on poisonous fruit to no ill effect? Were natural drugs already in use before humans appeared in the world? Currently, these tantalizing questions have no final answers, but it is certainly conceivable that natural medications were not a human discovery. We may be uncertain about which creatures deserve credit for first using natural drugs, but we can imagine that their earliest medications were substances ingested directly, comparable to the chimpanzee's "medication." Over time humans must have refined some medications into liquid extracts resembling those prepared by African and South American shamans today.

Although the prehistoric origins of the first human drugs are lost to us, we do have some more definite indications about early drug use in Europe. Until recently, the earliest firm record placed opium as a well-established drug in the Mediterranean nearly four thousand years ago. Now, new evidence suggests Europeans were already familiar with natural drugs more than a thousand years before that. This is one of the exciting findings that has emerged from extensive inquiry into a unique prehistoric event. In 1991, a man's body surfaced from an alpine glacier on the Austro-Italian frontier. The body is now widely known as the Iceman and is estimated to be about fifty-three hundred years old. Fortuitously dehydrated by the wind and then frozen shortly after death, the corpse had been preserved in glacial ice for more than five millennia. By the standards of his era, the Iceman was elderly, perhaps fifty years old, but his remains were in reasonably good condition and have afforded a remarkable record of his life and times. Teams of specialists have scrutinized his body, clothing, and possessions in minute

ILLUSTRATION 13 These bits of woody
material were well enough preserved after
five thousand years in ice to be identified as
a common alpine fungus.

detail. Their findings and inferences continue to educate us about
the Iceman's Stone Age world.

Among his possessions, the Iceman carried two walnut-sized
pieces of a tough substance threaded onto a leather thong. Initially
these objects appeared to be tinder, but it is now clear that they are
pieces of a woody brown-rot fungus (*Piptoporus betulinis*) that is a
familiar parasite on birch trees. This fungus causes brief diarrhea
when ingested and it contains compounds that are toxic to some
parasites. Could the Iceman have carried the fungus as a medication? Considerable support for this idea came from an autopsy revealing that the Iceman had suffered from whipworms (*Trichuris
trichiura*), which are very common parasitic nematodes. Whipworms cause acute stomach pain and can also lead to anemia, and,

indeed, the Iceman had signs of anemia. Anthropologists now infer that the Iceman was probably aware of his condition and consumed bits of the fungus to relieve his distress. Each dose should have killed a portion of his worms and purged a large number of them and their eggs from his intestine. This birch fungus may have been the only medication for whipworms available in Europe during the Iceman's lifetime.

Though better treatments have long since displaced birch fungus, opium and some other medications that were in use perhaps four thousand years ago have survived into our times. One of the most remarkable of these was already acclaimed in the Middle Kingdom of ancient Egypt, and yet its modern equivalent remains a treatment of choice in Western medicine today. This long-lived drug came originally from a plant known as bishop's flower (*Ammi majus*), a common Old World weed that grows extensively in the Nile valley.

Ancient Egyptian physicians used the juice or an extract of bishop's flower to treat a condition known as vitiligo, a pigmentation disorder characterized by smooth white patches of pigment-free skin. It can be brutally disfiguring and psychologically devastating, particularly in dark-skinned people. Egyptian practitioners rubbed the weed's juice on their patients' patches of vitiligo. Afterward, they had the patients lie in the intense Egyptian sun with the treated areas laid bare. With many applications over time, the whitish patches gradually darkened with pigment and became less prominent.

At about the same time far to the east in India, physicians were prescribing a strikingly similar regimen for the same condition. Their medication for vitiligo was an infusion of the seeds and roots of a plant known locally as bavachee (*Psoralea corylifolia*). Patients drank the infusion or applied it directly to their skin. After treatment, Indian patients also exposed their patches of vitiligo to the sun.

For the Indians, vitiligo was a serious condition. The *Atharva Veda*, a sacred Hindu text written no less than thirty-four hundred years ago, contains an entreaty addressed to a plant "rich in color," beseeching it to stain dark the hated white splotches. Vitiligo has remained a problem in India into modern times. Popularly called white leprosy, it can cruelly diminish the quality of life of its victims, who have traditionally been social outcasts with little chance of marriage or employment.

Since beneficial treatment of vitiligo simply requires safely darkening the light patches, success or failure would have been obvious to all. Application of local plants in combination with sunlight must have brought results, or the therapy would have never survived. Interestingly, a strong link exists between these odd procedures and modern Western medicine. A widely employed contemporary therapy for several skin disorders traces its origin directly to the ancient use of bavachee and bishop's flower.

Modern scientists found several decades ago that both of these plants contain compounds called psoralens (a name taken from *Psoralea corylifolia*), which most likely are agents of chemical defense for the plants. With this finding in mind, and after carefully observing herb doctors at work in Cairo, an Egyptian dermatologist first treated vitiligo with a purified psoralen (a compound called methoxsalen) in the 1940s. His clinical success eventually persuaded medical investigators in Boston and Vienna to experiment with methoxsalen for treatment of skin disorders. Favorable findings led to rigorous clinical trials of the drug in the 1970s.

This work finally resulted in a new protocol for treating psoriasis, a chronic skin disease that can be seriously debilitating. The patient receives successive oral doses of methoxsalen followed by ultraviolet irradiation of the affected areas of skin. The procedure effectively controls psoriasis, and with improvements incorporated over the past twenty years it has become a standard therapy. Related procedures bring relief from several other skin diseases. In addition, the combination of psoralens and light has a key role in a promising treatment now under development for a white blood cell cancer.

Psoralens increase pigmentation and bring about other changes through a series of complicated steps. After ingestion or direct topical application, psoralen molecules initially collect near the skin surface. Here they absorb light and immediately become chemically highly reactive. The light-energized molecules rapidly attack nearby DNA, crippling its performance by forming new psoralen-DNA chemical bonds. More psoralen molecules react as more light is available, and this explains the need for sunlight or ultraviolet irradiation as part of the therapy. This attack on DNA sets off a cascade of events whose ultimate consequences include formation of new skin pigment and destruction of diseased skin cells.

Investigators now understand the details of this cascade, thanks to decades of work in both basic science and medicine. In appreciating the benefits these investigations have brought us, it is worth bearing in mind how research on psoralens began. The initial discovery came four thousand years ago, when Egyptian and Indian physicians found that certain plant extracts and bright sunlight were an effective treatment for vitiligo. In the 1930s, psoralens were part of an age-old folk remedy offered by herbalists on the streets of Cairo. Fifty years later, physicians had incorporated them into modern medicine. The development points to some interesting questions: How frequently do traditional medications gain acceptance by orthodox medicine? What is the connection between folk remedies and approved drugs? How do the two differ? Are remedies unaccepted in Western medicine nonetheless effective? Traditional treatments come from natural compounds, so perhaps we should look briefly at these matters.

The most apparent difference between traditional and orthodox medications is how their usefulness is determined. A traditional medication's reputation rests on use and history, stretching centuries or even millennia into the past. The favorable experience of earlier generations serves as an endorsement of safety and efficacy.

For an orthodox drug, on the other hand, assurance of safety and efficacy comes from carefully controlled clinical trials. The idea of testing drugs in a controlled fashion goes back to 1754, when a Scottish naval surgeon, James Lind, investigated scurvy in the Royal Navy. A condition now understood as a vitamin deficiency, scurvy is characterized by swollen joints, spongy bleeding gums, large bruises, and eventual death. On a long voyage, it could destroy half a ship's crew. In a clinical trial on twelve sailors ill with scurvy, Lind tested six daily medications: cider, oil of vitriol (concentrated sulfuric acid), vinegar, oranges and lemons, seawater, and a concoction of garlic, myrrh, radish, and balsam. Those receiving oranges and lemons were cured in six days, whereas the others remained ill. The results supported Lind's hypothesis that citrus fruit prevented and cured scurvy.

In a modern controlled trial, clinicians compare a candidate drug's performance with that of a placebo or dummy medication. To evaluate a new medication for treating, say, enlarged prostate or advanced breast cancer, they recruit appropriate volunteers suffering from the condition as subjects for the trial. Half these patients, chosen at random, receive the medication under scrutiny, and half receive an identical-looking placebo. The comparison employs a so-called double-blind test, in which no one, neither patients nor physicians, knows throughout the trial who is receiving which treatment. Only after the trial is complete are the coded medications identified as drug or placebo. In this way, no presumptions about how a patient should react to the treatment offered can influence the trial's outcome. Experience has shown that placebo controls and double-blind testing are absolutely necessary safeguards to avoid subtle bias and to obtain reliable, objective results.

Such testing is costly, accounting for about a third of the typical 360-million-dollar expense of bringing a new orthodox drug to market. One might object that expensive and extensive testing is superfluous for medications that have a long history of traditional use. How could the safety and efficacy of such treatments be questioned? For several reasons, clinical trials are informative even in these cases. Physicians learn about side effects, unfavorable interactions with other drugs, and complicating factors that may affect only certain subgroups of patients. They can standardize the dose and chemical composition of drugs under examination. Even a long history of use may not furnish such information. Many traditional preparations, for example, are unstandardized plant extracts that are often administered under conditions that are neither regulated nor scrutinized.

There is another highly significant reason for such placebo-controlled double-blind trials: They allow consideration of what is called the placebo effect. Generations of physicians have learned through experience that a patient's condition may improve simply because he believes he is receiving powerful medicine and good care. A patient convinced he should get better may well do so. The improvement is real but cannot be credited to medication. How this comes about is not clear in detail, but the placebo effect is currently under intense study. At the same time, there is serious interest not only in discovering the mechanism of the effect, but also in finding ways to make the effect clinically useful.

The placebo effect manifests itself in many ways, revealing that a host of bodily reactions can be influenced by mental attitude and by what an individual believes should happen. One widely noted Japanese study reported monumental effects on the allergic reactions of thirteen volunteers who were highly susceptible to poison ivy. An investigator rubbed one arm of each volunteer with a harmless green leaf, which he identified as poison ivy. He touched the volunteer's other arm with an authentic leaf of poison ivy, which he said was harmless. Despite the subjects' known susceptibility, only two of them showed an allergic reaction to the really poisonous leaf. All thirteen subjects, however, developed a rash where they had been rubbed with the misidentified leaf that was actually harmless.

As in this Japanese study, the placebo effect can also be extremely powerful in drug trials. It is normal for 15 to 20 percent of patients receiving dummy pills in a clinical trial to show varying degrees of improvement. The number can be greater than 35 or

40 percent. At the same time, some highly regarded drugs benefit no more than 20 or 30 percent of patients receiving them. Failing to take the placebo effect into consideration could lead to utterly erroneous evaluation of a medication.

How do traditional remedies fare in such trials? Some perform quite well and prove to be highly effective, but others are no better than placebos. One striking success is an extract of sweet wormwood (*Artemisia annua*), which Chinese physicians have prescribed for the chills and fevers of malaria for more than two thousand years. About twenty-five years ago, Chinese chemists obtained from sweet wormwood its principal active component, a compound now called artemisinin. Clinical trials on malaria patients in Southeast Asia agreed with Chinese tradition on the value of artemisinin and also identified a few even more useful drugs prepared from it in the laboratory. These compounds are effective against the deadliest form of malaria and are now frequently the therapies of choice for treating it.

Some experts have hailed artemisinin as the most significant antimalarial discovery since quinine, and quinine first reached Europe in the seventeenth century, following the Spanish conquest of Peru. The significance of this expert response is apparent when you realize that malaria is one of the world's most destructive diseases, killing one to three million people each year, most of them children in Africa. Artemisinin could become a spectacular weapon against this plague, but unfortunately, sweet wormwood contains rather little of this wonder drug. For artemisinin to fulfill its promise, there must be enough available for public health agencies to dispense it widely for years. The present supply could not meet this demand, and research aimed at increasing it is under way. Some scientists are working to find an equally effective man-made compound related to artemisinin. Others are investigating various ways of boosting the natural production of artemisinin in the plant. If successful, this research could have a profound effect on public health in the developing world.

Another success story from Chinese medicine is an extract from the ginkgo tree (*Ginkgo biloba*) known as EGb 761. Ginkgoes are the sole survivors of a once large group of plants that flourished long before the appearance of flowering plants 135 million years ago. The tree is native to China, where for millennia its extracts have been employed for conditions involving the heart, lung, and brain. EGb 761 is now widely used in Europe for cognitive disorders and was recently approved in Germany for treating dementia such as Alzheimer's disease. Ginkgo extract has become one of the most popular alternative medications in the United States, particularly for memory enhancement. According to *Nutrition Business Journal*, sales in 1997 amounted to $240 million.

Here again, a placebo-controlled, double-blind trial found EGb 761 safe and effective in stabilizing patients with Alzheimer's disease or other dementia for six to twelve months. There were no unfavorable side effects and no harmful interactions with other drugs. In this case, physicians evaluated the extract itself rather than a purified compound, because EGb 761's action depends on several components. On completion of the trial, a relative assessed changes observed in each patient over the year. On average, these changes were modest. Patients with mild to moderate dementia who received EGb 761 had improved somewhat, whereas those on the placebo had deteriorated to a similar extent. There seems to be general agreement now that ginkgo is somewhat helpful for these patients. These trials do not address the possibility that it might also enhance memory in healthy people.

Another classic Chinese drug widely employed in the United States is dong quai, the crushed dried root of Chinese angelica (*Angelica sinensis*). With a long history in Chinese medicine as a woman's drug, dong quai is recommended for regulating the menstrual cycle, relieving menstrual pain, and treating problems associated with menopause. In the United States, its most frequent users are women suffering severe menopausal symptoms who elect not to undertake estrogen-replacement therapy.

Dong quai underwent a standard trial in seventy-one post-menopausal women with a history of night sweats or hot flashes. These tests failed to reveal any benefit from taking the medication. There were no significant differences in symptoms or physiological results between women receiving dong quai and those receiving a placebo. The California physicians who conducted these trials pointed out that their results do not speak to the value of the herb as it is traditionally employed. Chinese physicians always prescribe dong quai together with other herbs, and in the mixture it could be beneficial. In contrast, herbal remedies for menopause in the United States commonly consist of dong quai alone. Women who employ these remedies are apparently not receiving the traditional treatment and, according to this trial, their medication is ineffective. The investigators plan to examine a typical Chinese mixture of herbs in another group of menopausal women.

Remedies that fail to survive stringent examination are by no means limited to Chinese or other traditional sources. The United States has a long history of unorthodox medications, many of them based on plant extracts. Cancer and rheumatoid arthritis seem to attract especial attention, perhaps because orthodox therapies frequently offer little relief from these afflictions. In controlled testing, many widely touted novel medications perform no better than placebos. Proponents of such alternative treatments have often complained that their medications were unfairly denounced and that they were victimized by the medical establishment.

Perhaps this charge will be less credible in the future. American public interest in alternative medicine has expanded dramatically in recent years. According to polls, interest is particularly strong among persons of above average education who are in relatively poor health. The alternative medications they adopt are available largely as food supplements and herbal remedies; in 1997 the market for these commodities amounted to $3.6 billion. Orthodox medicine has not ignored this solid trend, and many American medical schools now offer courses and sponsor research into alternative therapies that were once ignored. At Congress's behest and with an annual appropriation increased from $20 to $50 million, the NIH recently broadened its own research efforts in this direction.

Unorthodox medications are likely to yield numerous truly useful drugs along with others that are no more effective than placebos, as more and more of these preparations are scrutinized. The pharmaceutical industry is currently enthusiastic about evaluating potential drugs from natural sources, and traditional pharmacopoeias are a rich source of leads. They have already yielded promising candidate drugs for treating HIV (the AIDS virus), leishmaniasis, and cancer, among other conditions. After evaluation, these candidates may reach orthodox medicine in the United States as medications approved by the Food and Drug Administration (FDA). Meanwhile, as untested medications grow in popularity it is worth keeping in mind that herbal medications, no less than drugs derived from other sources, are all chemical compounds. Regardless of their origins and history of use, chemical compounds may be medically useful, as well as dangerous to health if misused. There is no substitute for being well informed.

The pharmaceutical industry's current interest in naturally derived drugs also includes sources other than traditional medicine. In several programs now under way, drug houses have teamed with botanical gardens and academic institutions to examine organisms systematically for drug leads. Joint programs screen plants, and in some cases animals as well, from specific sources, searching for compounds with beneficial biological activity.

Only a tiny percentage of organisms have ever been screened for their content of potential drugs. The estimate is that worldwide only 5,000 flowering plants have undergone careful examination, out of more than 250,000 known species. The percentage of animals that have received attention is even smaller. The actual size of the total pool of species available for investigation is unknown, because no one has more than the roughest guess how many living species there are. Biologists have formally described about 1.4 million species, but undescribed ones number at least in the several millions. In any event, the pool of species is vast and the search for natural drugs could be virtually endless.

It is fortunate that there is an enormous supply of species, for only a very small fraction contain medically interesting chemicals. Each year about twenty thousand extracts from various parts of plants enter screening programs. The current system sponsored by the National Cancer Institute (NCI, a division of NIH) permits rapid testing of each extract against more than one hundred types of human cancer, as well as against HIV. In addition, drug houses conduct their own screening for many types of activity.

Most extracts are inactive in NCI's tests. Only about 2 percent of them merit further attention. From this 2 percent, only a tiny number of purified compounds is eventually selected for extensive assessment. Few of these candidate drugs ultimately survive the laboratory and clinical testing required for FDA approval. They fail most frequently owing to excessive toxicity and unacceptable side effects. The result is that since 1986 the screening of diverse extracts from forty thousand plant species has yielded only five compounds displaying significant activity against AIDS. The yield of drugs to fight cancer is no better. Since 1960, the number of plant-derived compounds approved as anticancer drugs is six.

Though screening and testing yield only a handful of drugs, these new medicines have a tremendous social and economic impact. Cancer chemotherapy alone saves about thirty thousand lives each year in the United States. More broadly, about 30 percent of the currently prescribed drugs in the United States are natural chemicals or compounds derived from them, and newly discovered natural drugs become available every year. As for the future, a 1995 economic study estimated a total of $3 billion to $4 billion as the value to the pharmaceutical industry of undiscovered natural drugs that may yet be developed through careful scrutiny of likely natural sources. The same study calculated that these prospective discoveries would be worth $147 billion to society as a whole.

Beyond the immeasurable human benefit, the overall economic value to the country also includes saved costs of additional

treatment, the value of work not missed, and similar considerations. The annual total value is estimated at $370 billion. Drugs
are, doubtless, the most significant application of natural compounds in the modern world, in terms of both their contribution to
human welfare and their market value.

Despite their significance, we still do not know why chemical
compounds from nature are effective against human disease. The
fact that occasionally they turn out to be good drugs is fortunate for
us humans, but it remains scientifically uncertain why this is so.
Why should natural compounds evolved for other creatures' defense serve human needs totally foreign to their original purposes?
The answer to this riddle is still unclear. Some scientists believe that
chemical strategies are restricted in nature, that there are only a
limited number of ways on a molecular level to attack or defend an
organism. This concept is related to the increasingly accepted idea
that evolution is conservative, making do and finding new uses for
what is at hand rather than undertaking grand new schemes. A
definitive answer to the question probably awaits an improved biochemical understanding of how both drugs and defensive compounds work. We have some details here but need more. This
uncertainty should not prevent us from looking at a few naturally
derived medications in more detail.

In the United States and other wealthy countries where many
people overeat, a huge market exists for drugs that safely reduce
cholesterol levels. Cholesterol is a necessary chemical in the human
body and we all synthesize it in our bodies. For some people, however, too much dietary cholesterol leads to excessively high levels
circulating in the blood. As everyone has learned over the past two
decades, high cholesterol levels can lead to clogged arteries, heart
attacks, and related unpleasant consequences. Lamentably, many
delectable foods—eggs, heavy cream, well-marbled roast beef, and
other fatty delights—provide more cholesterol than some people
can handle. Most people can maintain a safe cholesterol level by
attention to their diet, but some fail to do so. For others, even eating
wisely may fail to keep cholesterol at a healthful level.

The result has been a clinical demand for safe means of lowering cholesterol levels, a demand pharmaceutical houses have been quick to meet. Several effective medications are on the market, all of them derived from fungi. First to reach the market was lovastatin (marketed by Merck as Mevacor®), which is obtained from a mold called *Aspergillus terrus*. Soon another fungal derivative appeared, pravastatin (Bristol-Meyers Squibb's Pravachol®), and other drugs followed. When taken daily, each of the three or four leading compounds lowers cholesterol in the blood by about 35 percent, at a wholesale cost of about two dollars per dose. In 1996, the American market, which accounts for approximately half the global sales of these medications, amounted to three billion dollars.

Despite this success with new drugs to control cholesterol, medications from fungi are relatively unusual. Flowering plants provide more than one hundred of the currently prescribed pharmaceuticals and, since prehistoric times, have been the major source of natural medications. One of the six anticancer agents approved since 1960 is a recently introduced drug obtained indirectly from a common flowering plant. This potent medication, called etoposide, is now a first-line treatment for one type of lung cancer and is also prescribed for drug-resistant testicular cancer. Etoposide is not itself a natural product but a laboratory-prepared relative of podophyllotoxin, one of several poisonous substances found in mayapples (*Podophyllum pelatum*). Two other mayapple toxins are still under investigation as anticancer drugs. Podophyllotoxin itself proved to be effective against cancer, but clinical experience quickly revealed serious side effects, from nausea, diarrhea, and mouth ulcers to serious kidney damage.

Podophyllotoxin presented a familiar dilemma: what to do with an effective candidate drug that has unacceptable side effects. Confronted with this situation, investigators normally look for ways to suppress the unwanted effects and salvage the drug. They may combine it with other drugs or otherwise alter the way it is administered. They may also make simple changes in the drug's chemical structure. Chemists tried a number of structural changes on podophyllotoxin, and one set of these led to etoposide. The new compound retained podophyllotoxin's toxicity toward cancer cells but had no disqualifying side effects. Etoposide went on to survive clinical trials and take its place as a clinical drug to combat cancer.

Etoposide's parent, podophyllotoxin, comes largely from mayapple roots and rhizomes (horizontal underground stems from which roots grow), although all parts of the plant are toxic except the acidic yellow fruit. The mayapple is a small woodland perennial widespread in eastern North America, with a long history of medicinal use by American Indians. From the plant's dried and powdered rhizomes, tribes throughout the plant's range prepared an extract that was an effective laxative and also a purge for intestinal worms. The extract also found use as a poultice for skin tumors—a striking application, given the later development of etoposide as an anticancer drug. In early New England, neighborly Indians instructed colonists in the properties of mayapple extract. Later, others did the same for settlers who migrated west across the Appalachians. The immigrant Americans found the plant beneficial in both human and veterinary medicine. From such unpretentious beginnings, mayapples have given us a vital drug. In the United States alone, some sixty-five thousand gravely ill cancer victims benefit from etoposide each year. The drug has annual worldwide sales of about a half-billion dollars.

Plants have provided most of the natural medications now stocking the medical arsenal, but a new source of drugs may soon challenge this preeminence. Marine invertebrates—mollusks, sea stars, sponges, and many less familiar creatures—now provide a quarter of the new natural products tested in NCI laboratories. Although plants have been part of human life for thousands of years, we know almost nothing about most of these oceanic creatures. From marine bacteria to giant squids, the seas are filled with uncounted species whose chemicals we have never even imagined. Some of these species, such as jellyfish and sea slugs, have soft bodies; others, including corals and sponges, are stationary or too slow-moving to escape predators. Not surprisingly, many of the apparently defenseless creatures protect themselves with toxic chemicals. Compounds that discourage prowling carnivores may also destroy cancer cells, and so these organisms are attractive targets for drug screening. Some of these same organisms have long been a source of medications for traditional peoples.

One of the first promising marine products discovered through screening is a compound named didemnin B, produced by a Caribbean tunicate or sea squirt, *Trididemnum solidum*. These tiny creatures grow in colonies as a flat gray-green coating on coral reefs and underwater rocks. As an expanding colony overgrows corals, it eventually kills them, displacing the many other organisms that had made their home in the living reef. For this reason, these tunicates were generally considered undesirable until they became a potential drug source.

Considerable effort has gone into investigating compounds from tunicates over the past two decades. For unknown reasons these chemicals are often potent antiviral agents, whereas clinicians have few drugs active against viruses. Didemnin B was the first of these candidate compounds to be examined and initially it showed promise against a broad spectrum of viruses. After lengthy clinical trials, however, it was finally abandoned as too toxic for safe human use.

In the meantime, further screening of tunicates has yielded several other substances that are still under evaluation, including a very attractive compound called ecteinascidin 743 (ET 743). This candidate drug is from a different species of tunicate, *Ecteinascidia turbinata*, which colonizes the roots of mangrove trees in the warm coastal waters of Caribbean islands. ET 743 is not an antiviral drug, but clinical investigators report that it may be a powerful anti-cancer agent that some tests have shown to be more effective than are current drugs. More extensive testing is yet to come.

Other little-known marine invertebrates have also proved their worth to drug prospectors. One group of these is bryozoans or "moss animals," very small, gregarious organisms whose colonies encrust rocks, shells, and other surfaces, where they are often mistaken for corals or seaweeds. Bryozoans frequently become costly nuisances by colonizing ship hulls and fouling piers and docks. Nonetheless, a common bryozoan from off the California coast, *Begula neritina*, is the source of a very active antileukemia agent known as bryostatin I. Tests against other types of cancer are also under way.

Another kind of unfamiliar invertebrate that is exciting for drug investigators is sea hares, shell-less marine mollusks related to snails and sea slugs. Like many of their relatives that have given up their shells, sea hares safeguard themselves from predation by means of formidable chemical defenses. One well-defended creature is *Dolabella auricularia*, a hand-sized sea hare widely distributed in shallow tropical waters of the Pacific and Indian Oceans. This creature is the source of dolastatin-10, another highly active anticancer compound. Here early studies have been auspicious, and the compound is moving into advanced clinical trials against colorectal cancer and lymphoma.

Biomedical interest in exotic creatures raises serious problems of supply and demand. Organisms rarely need to synthesize their defensive compounds in large amounts. Often the threat of a potent defense is sufficient to ensure safety. A ton of tunicates, for example, yields only one gram (one twenty-eighth of an ounce) of ET 743, and many compounds of interest are much scarcer. Collecting useful quantities of scarce compounds can have a high price. If the organisms are rare or difficult to collect, accruing sufficient material even for clinical trials is arduous and costly. If the organisms are more easily gathered, there is the risk of disastrous overcollecting. Severely depleted populations can endanger the survival of a species, whose reproduction and development may be little understood. Depletion of populations of course also jeopardizes future supplies of the desired compound.

In principle, this problem has chemical solutions, such as laboratory synthesis of the natural compound or of some equally active substitute. However, such remedies are not always realistic possibilities. Synthesis may be too costly and substitutes cannot always be found. A solution that has been successfully adopted for plant-derived drugs is cultivation of the source as a crop. Fields planted to Madagascar periwinkles (*Catharanthus roseus*) in several countries provide stable supplies of vincristine and vinblastine, two widely prescribed anticancer drugs. Extending this idea to cultivating and harvesting marine animals for their drugs is now under investigation. With NCI support, teams of scientists are exploring techniques of farming in the sea, or aquaculture.

Aquaculture is simple enough for adaptable, easily grown organisms. In both Israel and Australia common green algae (*Dunaliella* species) are raised commercially. Thriving in intense sunlight and a wide range of salt concentrations, *Dunaliella* is ideal for cultivation in desert areas where there is brackish water or open sea. In such settings, the threat of loss of algae to predators is minimal. The algae are rich in beta-carotene, a highly marketable dietary supplement that is an antioxidant and a biological precursor of vitamin A. Under commercial growth conditions, the dried, harvested alga contains 4-6 percent beta-carotene by weight.

If exotic organisms can be cultured on a large scale as readily as *Dunaliella*, aquaculture could offer a stable source of drugs without raising the threat of species extinction. However, little-known organisms are often difficult to maintain in cultivation, and much research may be necessary before an invertebrate of interest can be turned into a farm animal. One such aquaculture project is currently under way in New Zealand. Here, the goal is to cultivate a rare local sponge (a new species of *Lissodendoryx*) that is the source of halichondrin B, a potentially valuable anticancer compound. Laboratory synthesis of halichondrin B on a scale necessary for clinical trials is impractical, as is a search for alternative synthetic substances. Successful sponge-farming seems to be the only practical way to overcome the obstacle of limited supply. The reluctance of sponges to thrive in laboratory settings elsewhere, however, suggests that such cultivation will not be easy.

It is not only marine invertebrates that now furnish new biologically active compounds to drug prospectors, but other creatures of the most diverse sorts as well. One biotechnology company hopes to develop a common hookworm protein as a clinically valuable drug to block blood coagulation. There is a promising new anticancer substance from sharks. A small amphibian known to Ecuadorian Indians as the poison-arrow frog has yielded a nonaddictive analgesic two hundred times more effective than morphine. Side effects prevented clinical adoption of this painkiller, but its chemical structure inspired the laboratory synthesis of a related compound that shows similar impressive activity without unacceptable side effects. Spiders, millipedes, insects, bacteria—all these and more are potential sources of beneficial drugs. There is enough to do to keep an army of scientists at work for decades.

Loose Ends and New Beginnings

From looking at special chemicals as drugs and other economically significant products, we return to chemicals in their natural settings. All across the biological spectrum creatures feed and flee, prey and are preyed upon, and seek and identify mates. For these and an inconceivable number of other activities, chemicals are frequently part of the story—carrying messages, supporting a lifestyle, providing the means of attack or defense. Earlier chapters presented activities where the role of chemicals is well known and the compounds themselves have occasionally been identified. Of course, countless other activities exist where a role for chemicals is an unsettled question. This chapter presents a few topics of this sort where investigation is still incomplete. These, then, are topics for current research, demonstrating the range of questions and possibilities that investigators face today. In each case, enough is already known to encourage more attention. In addition to their scientific interest, some of these topics might also have practical importance if they were better understood.

First is an exceptional relationship between yucca moths and yucca plants, two organisms that need each other to survive. This is a textbook example of obligatory interdependence that frequently appears in college courses on ecology and evolution. (Another relationship of this sort is that between fig plants and the fig wasps that pollinate them; virtually every species of fig is pollinated by its own species of wasp.) The relation between these moths and yucca plants was first discovered in a small American plant known as Adam's needle (*Yucca filamentosa*) and its moth (*Tegeticula yuccasella*). Subsequently, analogous associations between other moths and yucca species have come to light, particularly in the deserts of the American Southwest where yuccas abound. There is not a precise one-to-one relationship between these plants and insects. A yucca species may support several different moths and a particular moth species may associate with a variety of yuccas. All yuccas belong to the same genus (*Yucca*) whereas yucca moths represent three genera in a single family (Prodoxidae) of Lepidoptera.

The essence of the moth-yucca association is that yuccas depend exclusively on their moths for pollination, whereas the moths depend exclusively on the plants for egg sites and larval food. In more detail, their association works like this. At the end of pupation, adult yucca moths emerge from underground cocoons, fly to nearby yucca flowers, and mate. The female moth carefully gathers pollen from the flower's anthers using her distinctive bristlelike "tentacles," which are specialized mouthparts not found in other insects. Tucking the pollen between her tentacles and her thorax (more or less under her chin), she flies off to another, newly opened yucca flower. Here she deposits one egg or more in the flower's ovary after penetrating it with her hollow needlelike ovipositor, or egg-laying organ. Then she climbs to the flower's stigma, where with apparent deliberation she deposits some of the pollen she has brought, pollinating the flower. The moth flies off to find another yucca flower and repeat her tasks.

After pollination, seeds begin to form within the flower's ovary. Meanwhile, each moth egg deep inside the ovary hatches to a minute larva. The larva consumes several yucca seeds to support its growth but leaves behind many more that can go on to propagate the plant. Once a larva matures, it makes its way out of the flower and drops to the ground. After burying itself near the yucca, it spins its cocoon and pupates, later to emerge as an adult moth.

Both moth and plant are well adapted to their lifestyles. Because most yuccas produce little nectar, they offer no reward to attract other insects but entrust their pollination solely to yucca moths. The lack of nectar in yucca flowers means nothing to the moths. They have an incomplete gut and cannot feed in any event. Without nourishment, they live only long enough to fulfill their reproductive tasks. Two to four days after emerging from their cocoons, they die.

Biologists have known of these intertwined lives for more than a century, but only more recently have they brought modern tools to bear on the relationship. Along with several other points, the role of chemical signals here is still unsettled. The moths may locate yucca flowers by scent, sight, or both. For such a scientifically well-known interaction, it would be interesting to know more. If yuccas do emit a chemical attractant, the scent apparently attracts only yucca moths, a selectivity suggesting an attractant that mimics some unsuspected moth pheromone.

From moths and yuccas we turn to some beetles with odd eating habits. In tropical Central America certain beetles (*Leistotrophus versicolor*) spend much of their time around piles of dung or carrion where flies congregate. Both flies and beetles come searching for a meal. Many tropical flies nourish themselves on decaying organic matter and track their food by odor. In addition, chemical sensors in their feet alert them to a food source underfoot. They are well equipped to exploit a dunghill. As for the beetles, the dung interests them only as bait, for they have come to feed on the flies. They sit quietly and wait for inattentive flies crawling and feeding in the dung to blunder into their jaws.

These sites attract enough flies that beetles are assured of a satisfying meal. But what if the sites themselves are in short supply? Dung and carrion are in strong demand among local consumers and do not long persist on the forest floor. When they are absent, the beetles must look elsewhere for food. Showing considerable flexibility in their habits, they alter their hunting technique and re-place their sit-and-wait strategy with a more active one. Hungry beetles now frequent streams and rivulets where small flies gather in large numbers. In this setting, a beetle puts out bait by wiping its abdomen back and forth on a rock or leaf, apparently laying down a chemical attractant. Then it lies immobile, abdomen curled up-ward and antennae stretched forward, to await an inquisitive fly. When one lands nearby, attracted by the bait, the beetle slowly turns to face its victim, raising its abdomen in an arc over its body with the tip pointing at the fly. The abdomen's tip may quiver, and sometimes minute bubbles of fluid appear. Whatever the beetle is doing, the performance somehow mesmerizes the fly. A fly rarely flees as the beetle inches closer or even kicks it. Approaching cau-tiously, the beetle finally lunges forward and seizes the startled fly with its forelegs.

Beetles attending their riverside flytraps inevitably draw the attention of other foraging insects. Bees, wasps, and butterflies ap-proach them, sometimes putting forth their antennae to touch them. Perhaps the beetles' appearance is enticing, because an im-mobile beetle closely resembles the bird droppings that are a staple of many insects' diets. The beetles' chemical bait may draw some insects as well. Disturbing a hunting beetle this way might seem dangerous, but the insects' curiosity never costs them their lives. Although a beetle may grasp curious bees or butterflies that come sufficiently close, it always releases them, rejecting them as prey. The beetles refrain from eating these inquisitive visitors, yet they are not at all particular about the flies they eat, considering equally acceptable flies of many different shapes and sizes from many dif-ferent families. They snap up fruit flies measuring no more than 2 millimeters (less than a tenth of an inch) in length, as well as bottle flies and blow flies more than ten times larger. These flies appear to have little in common, apart from possessing two wings rather than the four of most other insects. Why do the beetles spe-cialize in flies as food? What has made them so selective and how do they make their choices? Do flies as a group possess chemicals that influence beetles, or is it some other attribute? These are mysteries yet to be explored.

Other specialist feeders include certain small larval insects that spend their time in termite galleries. The larvae (*Lomamyia latipennis*) belong to the order of insects known as Neuroptera, which also includes the green lacewings encountered earlier. Like aphid-eating lacewing larvae, these larvae are predatory carnivores, but they specialize in feeding on western subterranean termites (*Reticulitermes hesperus*), a species common in California and other parts of the western United States.

Female *Lomamyia* lay their eggs clustered on little stalks they attach to termite-infested wood. How they choose egg sites is unexplored. When the eggs hatch, the larvae crawl down the stalk and into the wood to search out termite nests, where they make their home. The resident termites seem to accept them, although the larvae are cautious when termites are nearby. Do the larvae smell like termites? Despite the apparent peace between them, the larvae live by preying upon the termites in an odd manner.

The youngest larvae weigh less than a small grain of salt but do not hesitate to take on termite workers thirty times their size. A tiny larva repeatedly approaches a termite and then retreats. Eventually, it maneuvers into a position with the tip of its abdomen pointing at the termite's head. The larva lifts the tip and waves it in front of the termite. This is its way of launching a deadly gas attack. Though termites are blind, they are extremely sensitive to nearby movement and defend themselves vigorously when assaulted. Nonetheless, the termite seems unconcerned by the larva's performance and makes no attempt to turn away or escape.

A few minutes after the larva's tip-waving, the termite collapses on its back, its legs twitching irregularly in the air. Within another minute or two, it stops moving and remains completely paralyzed over the next two or three hours until it dies. As the toxin takes effect, the predatory larva continues bobbing back and forth. Once the termite is motionless, the larva sets upon its prey, piercing and draining the termite's body with its sucking mouthparts. Older larvae feed similarly, although they do not bother waving their abdomen at a single termite. Being much better armed than their smaller comrades, one of them can immobilize as many as six termites in a single gassing.

Poison-gas-wielding larvae are strange, but there are things stranger still. The effect of the larval toxin is curiously restricted. It paralyzes the larvae's prey but not the larvae themselves. Moreover, it fails to paralyze various flies, wasps, and other insects, as well as the two other termite species native to the same part of California. These other termites may be unaffected simply because they are larger, but the insensitivity of other insects is unanticipated.

This selective toxicity of the larvae's weapon suggests practical applications. Western subterranean termites are economically significant pests, considered to be the most destructive of all California termites. A pesticide directed specifically against them and harmless to other creatures would be environmentally friendly and extremely useful. Exterminators could rid a house of termites without endangering the residents or destroying beneficial insects. The chemical nature of this toxin is still unknown. We can only wonder what substance can rapidly paralyze one kind of termite and leave other insects unscathed.

If we wish to indicate directions of present and future research, we must include some creatures from the sea. Seawater covers more than two-thirds of the earth's surface and accounts for a much larger fraction of the biosphere (the part of the world that supports life). Yet, in comparison with investigations on land, probing the oceans is more difficult and appropriate tools are less well developed. It is no surprise then that marine life offers the last great frontier for biological exploration. As the ocean frontier expands and the secrets of its depths unfold, novel discoveries continue to appear. It was news in 1999 when an unmanned Japanese deep-sea vessel photographed fish at much greater depths than previously recorded. Some years earlier, hydrothermal vents and their hitherto unimagined forms of life first suggested new scenarios for the origin of life and raised a host of issues for research in fields from biochemistry to evolution. At about the same time, samplings of organisms from the deep sea floor revealed an unforeseen richness and variety of creatures populating a hidden world once thought to be barren. As marine biology matures, so the chemical ecology of marine organisms has been an expanding subject for more than a decade. This is a rich area for future discoveries.

Our representative marine creatures are sea hares, which are an order (Anaspidea) of mollusks common in the coastal waters of many parts of the world. They resemble large shell-less snails, inching along the bottom and feeding on seaweeds. Some are about 12 centimeters (5 inches) long, whereas larger species may be three or four times that size. Two tentacles on their heads look something like small rabbit ears and give them their odd name. Though sea hares lack a protective shell, they have few enemies. Some sea anemones prey on them, but most other animals leave them alone.

Predators ignore sea hares because, like other mollusks without shells, they have chemicals that afford them the protection once given by a shell. (We saw earlier that a compound from a tropical sea hare is now a promising anticancer drug.) Many sea hares have an acrid or fetid odor that is distinctly unpleasant to humans. Their egg masses and their skin must be distasteful, because one exploratory bite is enough to convince a would-be assailant to look elsewhere for food. For some species, the deterrents responsible for these properties may come directly from their seaweed diet, but other species seem to synthesize these defenses for themselves. The evidence is mixed and confusing.

To supplement these odor and taste defenses, sea hares have two chemical responses they can deploy against an active aggressor. When frightened or aggravated, they are quick to release a jet of ink. A minor threat rates only a small jet, and a severe disturbance elicits much more. The ink contains a protein about which nothing is known, along with seaweed-derived pigments. Because sea hares incorporate these pigments into their ink, many that feed regularly on red algae have ink that is purple in color. Ink has a

disagreeable taste and probably serves mainly to discourage potential predators. It may also signal imminent danger to other nearby
sea hares. There are bits of information about the ink from various
species, but a coherent chemical account has yet to emerge.

There is also another defensive secretion that sea hares release
less readily than ink. This is a milky white liquid called opaline, so
viscous that it can be stretched out through the air into a long
string. For sea hares, opaline is what is known as a defining characteristic, because all sea hares emit opaline whereas no other species
are known to do so. In spite of its defining role, opaline remains
poorly understood. It contains proteins and perhaps components
derived from the animals' diet; the evidence here is in dispute. A
sea hare releases opaline less readily than ink but seems to do so as
a second response to serious predatory assaults. When touched by
a sea anemone's tentacles, a sea hare discharges a shot of opaline,
which causes the anemone's tentacles to contract. No one yet
understands how opaline deters predators or what chemicals are
responsible for its activity.

Sea hares are relatively common animals. They have large and
readily accessible neurons, and so one easily available genus (*Aply-
sia*) is regularly studied by neuroscientists who are interested in
learning and memory. In contrast to the wealth of neurobiological
information from their research, knowledge of sea hares' well-
developed chemical defense systems is surprisingly meager.

Finally, we turn to an astonishingly complex arrangement
among a flower, a butterfly, an ant, and a wasp that features both
a parasite and a parasitoid. Without the merging of these four organisms' lifestyles, neither the butterfly nor the wasp could live. As
it is, the butterfly and the wasp in this case are rare species. Chemicals probably enter this elaborate four-way connection at several
points, but very little is known for certain.

This association is best understood by following this particular
butterfly's development. This species (*Maculinea rebeli*) is one of
several closely related European butterflies known collectively as
large blues. It is one of the world's most specialized creatures,

being able to live only in a location where a handsome plant called cross-leafed gentian (*Gentiana cruciata*) thrives side by side with a particular species of ant (*Myrmica schencki*). The ant and plant flourish in grassy meadows and on hillsides, either together or separately. They seem to have nothing to do with each other. Because the butterfly can survive only where both are present, it exists only in isolated populations, often consisting of fewer than one hundred individuals. These little islands of butterflies can expand their range only into areas where both ant and plant have already established themselves, so the islands are rarely able to combine.

In summer, gravid female butterflies search out gentians in bloom and lay their eggs on the small blue flower heads. The eggs hatch in July or August, and for the next two or three weeks, minute caterpillars develop on the flowers. Then, while they are still quite small, the caterpillars drop from the plant to the ground, where several kinds of ants (all *Myrmica* species) may discover them. Presumably owing to the caterpillars' odor, foraging ants that come upon them mistake them for their own larvae. They promptly pick the caterpillars up and carry them back to their underground nests, there turning them over to attentive nurse ants. The nurses rule over the brood chamber, where they have custody of the colony's larvae and pupae. They gladly add the caterpillars to their nursery. Under their care, the caterpillars increase in weight more than one hundredfold over the next ten months, from 1–2 milligrams (the weight of about four grains of salt) at adoption to nearly 200 milligrams by the time they pupate.

Not all caterpillars achieve this comfortable lifestyle. Enjoying the ants' hospitality depends on being taken home by the right kind of ants. The caterpillars have no control over this crucial matter, because the first *Myrmica* ant to notice a caterpillar takes it off to its own nest. Although each *Myrmica* species can distinguish its own larvae from those of related species, all local *Myrmica* ants accept the caterpillars as their own. How this comes about is unclear. Like other insect larvae living openly in ant nests, the caterpillars presumably carry the odor of ant larvae. Perhaps they smell enough like all the local *Myrmica* larvae to deceive each species in turn. If so, they routinely fool all the ants even though the ants' own larvae cannot. Fooling all the ants all the time might seem to be advantageous, because it increases a caterpillar's chances of acceptance into an ant nest. The difficulty is that these caterpillars can succeed only when taken in by one particular species of ant, *Myrmica schencki*. In other *Myrmica* nests, the caterpillars die—yet another mystery to be explained.

Overall, the odds are against the caterpillars. The right ants are likely to find a caterpillar only if their nest is within about 2 meters (6 feet) of the gentian plant where the caterpillar hatched. Even so, a different kind of ant from another nearby nest may remove the caterpillar first. It is also possible that no ant will come upon the caterpillar before it starves or is eaten. Making the situation worse, gravid female butterflies know nothing about ants. They lay their eggs in gentian flowers whether or not ant nests of any sort are nearby. Even before hatching, some caterpillars are fated to die. As might be expected, most of these caterpillars do not survive. Their survival depends on events that seem to have nothing to do with their own fitness.

The few caterpillars adopted into the correct ant colony enjoy a good life. Each of them has as many as five ants solicitously supplying it with regurgitated food. After ten months of ample feeding and little exertion, the impostors retire to pupate in the upper portion of the ant nest. Before reaching this stage, however, successfully adopted caterpillars face still another obstacle. Wasps are on the prowl for them, and once again their fate is largely a matter of chance. One species (*Ichneumon eumerus*) of the many thousands of small parasitoid wasps seeks out these caterpillars as sites for its eggs. The wasps are as particular about egg sites as the caterpillars are about ants. Their larvae can develop only within the caterpillars of this single species of large blue butterflies. Because the caterpillars are rare, underground, and out of view, the wasps' search for egg sites might seem hopeless. Yet, gravid wasps know how to track down what their larvae need. They actively seek out these caterpillars that are concealed deep inside their underground sanctuaries.

The wasps' habit of penetrating subterranean ant nests to hunt down caterpillars is remarkable and dangerous. Female wasps' first task is to locate the proper kind of ant nests. Because nests with caterpillars are necessarily near gentian plants, it would not be surprising if the wasps concentrated their searches around the plants. Strangely, evolution of the wasps' lifestyle has taught them nothing about gentians. They look throughout the grassland whether gentians are close by or not, running in short spurts here and there and tapping the ground with their antennae. Their behavior changes when they come within about 1 centimeter (a half inch) of a foraging *Myrmica* worker. Suddenly, they walk more slowly, making frequent sharp turns and probing any crevice in the ground with antenna taps. If a wasp fails to discover an ant nest in ten to twenty minutes of carefully exploring the locale around a foraging ant, she flies off to another grassy patch and repeats the process.

From time to time, a wasp's probing pays off. She locates a small opening in the ground that marks the entrance to a *Myrmica* ant nest. If it is a nest of the species that might harbor suitable caterpillars, she begins a rapid drumming with her antennae on the ground around the nest entrance. If she determines there are caterpillars down in the nest, she may poke her head into the tubelike passage and enter it. It is astonishing that the wasps can detect whether the subterranean nest contains caterpillars. About 20 percent of them penetrate a nest holding caterpillars, but they almost never enter if caterpillars are absent. Possibly a third of those wasps entering a nest move on down the passage into the ants' brood chamber, where they sting a caterpillar and deposit an egg. (A wasp's hollow sting is also her ovipositor.) Overall, about 5–6 percent of the wasps that come upon an appropriate nest entrance complete the entire egg-laying sequence.

A wasp presumably identifies appropriate nests by odor, discriminating among very similar ant-recognition pheromones. Virtually the only chemical information available in this entire set of

ILLUSTRATION 14 Ignoring a nurse ant
attacking her leg, a female wasp grasps a
caterpillar and prepares to lay an egg.

relations is that the recognition pheromones of various *Myrmica* species differ only slightly from one species to another. They all are mixtures of the same components, combined by each species in somewhat different ratios. The difference from species to species is subtle, but from the surface the wasp recognizes the right ants' odor. If an unsuitable *Myrmica* species resides below, she rejects the site, moving away to resume her search elsewhere.

If a searching wasp very probably identifies suitable ant nests by odor, it is much less clear how she senses from the surface which suitable nests actually contain caterpillars. After all, the caterpillars have successfully disguised themselves by carrying an odor the ants accept as that of their own larvae. Is the wasp able to discern tiny differences between ant and caterpillar odors that the ants themselves fail to detect? Even if she is capable of this, it seems incredible that she could do so from outside the nest, where caterpillar odor should be swamped by the ant larvae's very similar odor.

Instead of relying on odor the wasps may detect caterpillars by sound. Both ants and caterpillars stridulate, or produce sounds, and they do so in similar rhythms. When amplified, their songs sound different to human ears, and maybe they do to wasps as well. Here again, the problem arises of detecting signals from caterpillars greatly outnumbered by ants. Could the wasp's hearing be so acute? The puzzle remains unsolved.

A wasp that detects caterpillars and moves down the entrance passage to the nest typically comes upon several young caterpillars in the brood chamber. At the time of the wasp's intrusion, the nurse ants have been feeding and cosseting them for about two weeks. Already, two groups of caterpillars can be distinguished. There is a fast-growing group now weighing about 40 grams apiece. These will likely survive and eventually pupate. A second group grows more slowly and weighs only about 8 grams at this point, and is almost certain to die before pupation.

Only the larger caterpillars interest the wasp. Her choice may be influenced by the nurses' frantic activity. On the wasp's arrival, they begin picking up the smaller caterpillars and removing them to safety. Also, it may be easier for the wasp to grasp and mount the larger caterpillars, which she must do to deposit her eggs. In any event, her choice secures a good chance of survival for her offspring. The larvae that hatch from her eggs remain within their caterpillar hosts through pupation. A caterpillar that hosts a wasp larva pupates like those that do not, but from its pupal case there emerges not an adult butterfly but an adult wasp.

The egg-laying wasp visits all the large caterpillars in the brood chamber, allotting a single egg to each one. In the laboratory, a wasp never returns to a caterpillar to lay a second egg. She marks each victim as she stings it, probably leaving a pheromone on its surface as she withdraws her sting. For several days after being marked, a caterpillar is immune to further attack by wasps. By placing only a single egg in each host caterpillar, the wasp gives her offspring another advantage. Two wasp larvae confined within a single host would have to compete for nourishment, and probably neither would survive.

As the wasp proceeds from caterpillar to caterpillar, depositing one egg after another, the nurse ants struggle to oust her from their brood chamber. They discover that the intruder in their midst is robust and not easily expelled. Even with twenty or thirty ants swarming over her, she retains her hold on a caterpillar and lays her egg. She also deploys a very effective chemical weapon in fending off the ants. This is a spray that sows dissension and confusion among her foes, much like the chemical weapon some slave-making ants employ. When she sprays the ants, many of them abandon their effort to expel her. Instead, they break up into groups of four or five and begin fighting among themselves. Between these altercations among nestmates and the nurses' preoccupation with saving the brood, only a small fraction of the ants at hand are actively engaged with the wasp.

In this chaotic setting, the wasp completes her egg-laying chores. Only then does she turn to leave. As she starts making her way back up the passage, a host of angry ants accompany her. All the way to the exit, they bite her legs and wings to hasten her withdrawal. On reaching the ground surface, the wasp beats her wings repeatedly to shake off her assailants. Only after about ten minutes do the last of the ants give up and retreat to the nest. Despite their best efforts, they have failed to injure the wasp. Her wings are intact, undamaged by the repeated biting, and she has successfully placed a few of her eggs. Now, with no time for rest, she takes up her methodical search for caterpillars once again.

We can only marvel that such lifestyles ever arose and are successfully maintained. One significant factor in their continuation is that a large portion of caterpillars survive once they reach the nests of appropriate ants. A butterfly population can persist even when 95 percent of the caterpillars are adopted by the wrong ants and so are destined to die. The successfully adopted caterpillars' high survival rate also helps justify the wasps' risky behavior in laying its eggs. Putting eggs into caterpillars that are likely to pupate favors survival of the wasps' descendants as well. If a wasp can lay five to ten eggs in one or two ant nests, there is a good chance that at least two of her progeny will survive.

A number of special chemicals presumably contribute to the success of these creatures: defensive compounds, marking pheromones, recognition signals of both ants and caterpillars (and perhaps of gentians and wasps as well), and maybe some other as yet unrecognized possibilities. Learning about the chemicals important in these intertwined lives should be fascinating but demanding, as studying the activities of such rare creatures is experimentally quite difficult.

The topics of this chapter provide a window into research in special chemicals and are an indication of the myriad chemicals remaining to be probed. Worth noting is that the sampling of current knowledge we have explored is the direct result of earlier probing into questions of this sort. It is evident from what we have seen here that there are still rich sources to mine for future discoveries and practical applications.

Complexity in the Real World

11

In our exploration of living organisms' special chemicals we have focused on the immediate effects of chemical contacts. A seed broadcasts signals that encourage an ant first to pick it up and then to carry it home. Defensive chemicals in zebra odor repel tsetse flies and protect zebras from the flies' infectious bites. A house fly's sex pheromone brings male and female flies together for mating. Most of what we know about the effects of nature's special chemicals concerns such fundamentally important direct responses. Even the reports in the last chapter involving three and four species concerned mainly direct interactions between them two at a time. However, it would be an extreme oversimplification to think of the biological world only as a collection of species communicating with their own kind and making simple, individual connections with other species. The reality is that one- and two-species contacts often vitally affect another species or even many other species. We must now look beyond primary effects at the indirect consequences of organisms' contacts.

A nice example of these consequences comes from an extended investigation of a natural marine community on the Washington Pacific coast in the 1960s. This study does not concern chemicals, but it does illustrate dramatically how interaction of a few species

ILLUSTRATION 15 Along the Pacific
coast of North America, barnacles and
mussels are the ochre starfish's primary prey.

can alter other organisms' lives. In the rocky intertidal setting
under examination, the principal residents were California mussels
(*Mytilus californianus*), several kinds of barnacles (mainly an acorn
barnacle, *Balanus glandula*, with smaller numbers of several other
species), and common ochre starfishes (*Pisaster ochraceus*). Seaweeds, chitons, limpets, and a few other marine organisms were
also present, making a total of fifteen species. The starfish were the
principal predators in the community and fed on masses of the
various kinds of barnacles and some of the mussels.

The investigation concerned a stretch of shore divided into two
portions. One portion was kept free of starfish by frequently removing them; the other was left in its natural state. Observation of
the two shore areas over the next two years revealed how the absence of starfish transformed one of the two intertidal communities,

comparing its natural development with that of the unmodified area. At first glance it might appear that removing the predators would affect only their prey, and that without starfish the mussels and barnacles would now prosper. Alternatively, perhaps the predators' absence would benefit the entire community, enabling all the remaining fourteen species to thrive and increase. The actual outcome was quite different from these prospects. Exclusion of the starfish eventually reduced the remaining community from fourteen species to eight. Without the principal predator, the original complex system became unstable. As soon as there were no starfish to feed on the barnacles and limit their number, the barnacles began to spread, taking over more space and crowding out other species. The starfish-free system developed further with time and, finally, was dominated by mussels and a single, previously minor species of barnacle, the two organisms able to occupy the available space most efficiently. Owing either to lack of space or of food, many of the other species had disappeared. Even though most of them had nothing to do with the starfish's predation on mussels and barnacles, their survival became impossible without these interactions.

Long before the study detailing the role of starfish in this system, the intricacy and subtlety of extended connections were well recognized. For more than a century, naturalists have described the interdependence of organisms, marveling at the complexity of nature and the remarkable intertwining of different species' lives. John Muir, the renowned champion of the California wilderness, in 1911 remarked in *My First Summer in the Sierra* that "when we try to pick out anything by itself, we find it hitched to everything else in the universe." There are various ways of "hitching" living things together, and the one that concerns us particularly depends upon chemical interactions. Throughout the living world countless chemicals tie creatures to one another, often giving rise to effects felt elsewhere in their biological community. It is to these broader effects of chemical interactions that we now turn.

We can see these effects by examining a well-understood group of interacting creatures and the results of their chemical connections. Even partially understood ecological systems that provide information of this sort are rare, but a recent exhaustive investigation affords a fascinating example. This was a study designed to answer an apparently straightforward question: Why does the population of deer ticks in northeastern American forests fluctuate widely from year to year?

Asking about ticks was not a trivial question. (Ticks are larger versions of mites and, like them, are eight-legged parasites.) Deer ticks (*Ixodes scapularis*) bite people walking in the woods or sometimes even as they sit in their own backyards. Many of these ticks carry *Borrelia burgdorferi*, the corkscrew-shaped bacterium that causes Lyme disease, and they can transmit this pathogen when they bite. Each year about ten thousand Americans contract Lyme disease in this way, making it the nation's most common disease carried by insects or ticks. Thousands more become ill in other countries. Many victims suffer painful symptoms, and for some the disease has grave long-term consequences. In the United States, Lyme disease is particularly serious in the Northeast from Massachusetts to Maryland, and in Wisconsin and Minnesota, as well as in northern California and southern Oregon, where a related tick spreads the bacterium.

In seeking to control Lyme disease, public health officials have been curious about why there are such wide swings in the number of ticks from year to year. As the disease constitutes a more critical menace when the woods are particularly full of ticks, accurate predictions of heavy infestations would be advantageous. At one time, some specialists suspected weather patterns as a significant factor in controlling the tick population, but before this present investigation, no one really knew what was responsible. Even now, some ecologists disagree with the present conclusions, stressing that there are still uncertainties and loose ends to track down. In investigating variation in the tick population, biologists probed connections among a community of forest creatures that maintain a number of chemical interactions. Before examining these interactions, however, it will be helpful to introduce the community itself and the discoveries it yielded.

The question under study was simple enough for a child to grasp, but several years of research were necessary to track down its answer. It is not the weather that regulates the ticks, nor is it any of several other variables put forward at one time or another. What regulates the tick population is simply the size of the annual acorn crop. White oaks (*Quercus alba*) are common in the northeastern United States, and every three or four years they drop an especially large number of acorns. Biologists call this mast fruiting and believe it is an evolved trait that enhances the trees' likelihood of successful reproduction. The idea is that with a large number of acorns there will be plenty for new oak trees even after predators have eaten their fill. Mast fruiting is common in oaks and a number of other trees. In white oaks it is ultimately responsible for an increase in the tick population, for the number of deer ticks soars the second summer after a bumper crop of acorns.

An amazing chain of biological events leads to this unsuspected association of acorns and ticks. Assembling the links in this chain called for extensive field study and laboratory experimentation, all of which was the work of ecologists in New York and Oregon. Their efforts finally implicated two mammals, the white-tailed deer (*Odocoileus virginianus*), and the white-footed mouse (*Peromyscus leucopus*) as the critical links between ticks and acorns. The deer is a handsome grayish to reddish brown animal with white patches, nervous and shy, and weighing about 150 pounds. It is common from southern Canada to Central America. The mouse rarely weighs more than an ounce and has a reddish brown back over a white belly and feet. It too can be found throughout most of North America, in a range that includes desert as well as forest habitats. Both deer and mice feed on the white oak's acorns, and both have intimate associations with deer ticks.

The relationships among these creatures work in this way. The deer carry adult ticks, which reproduce in the fall. After gorging themselves on deer blood, gravid female ticks drop from the deer to deposit their eggs in the leaf litter of the forest floor. The eggs overwinter where they were laid and hatch the following summer into minute larvae. These are active little creatures that waste no time in attaching themselves to mice or other small mammals for a blood meal. After feeding, larvae drop off their hosts and molt to a somewhat larger next stage, called nymphs. The nymphs, in turn, take blood from another mammal, often an eastern chipmunk (*Tamias striatus*). On completing this blood meal, nymphs also abandon their hosts and molt, this time to adult ticks. The entire process from larva to adult extends over a year, so it is late in the second summer after the eggs were laid that newly matured adults attach themselves to passing deer and thus complete their life cycle.

If either adult or nymphal deer ticks choose a human host rather than their customary wild ones, their bites may lead to Lyme disease. White-footed mice are an excellent reservoir for Lyme bacteria, having infection rates as high as 90 percent in some areas. Infected mice transmit the bacteria in their blood to feeding larval ticks, which then develop into infected nymphs and adults. (With the exception of a small number that inherit the pathogen from their mother, larvae only become infected with Lyme bacteria during their single blood meal and so do not transmit the disease themselves.) While both nymphs and adult ticks can transmit Lyme disease to humans, it is nymph bites that are responsible for most infections simply because nymphs' small size allows them to escape human notice. On inspecting themselves for ticks, people are less likely to discover and remove the pinhead-sized nymphs than the larger adults.

Deer, mice, acorns, and ticks coexist in a natural balance that the unusually large acorn crop of a mast fruiting can alter. An exceptionally great number of acorns in the fall brings about a much increased population of mice the following summer, because only when acorns are plentiful do a large number of the mice have sufficient food to live through the winter. An ample acorn crop also draws more deer to the oak forest to feed. The deer bring their ticks

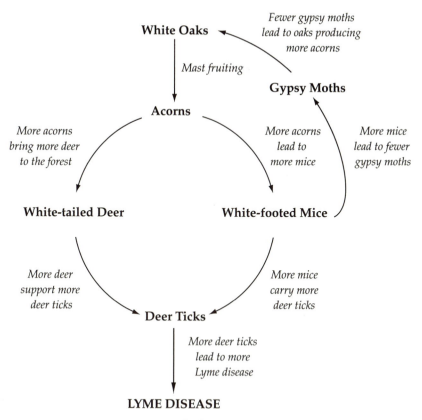

ILLUSTRATION 16 An unusually large acorn crop can result, two summers later, in an increased tick population and an increased incidence of Lyme disease.

with them, and of course more deer bring more ticks. This elevated tick population then lays more eggs than usual. An increased number of tick larvae hatch the following summer, when the forest is teeming with an unusually large number of mice. Finding more mice accessible to feed upon, more tick larvae secure their blood meal and more survive. Thus, as the ticks mature over the next year there are, first, more nymphs and then more adults than usual.

These are the main links between acorns and ticks. Each species naturally interacts with many other creatures in the forest, but most of these other interactions do not materially influence the tick

ILLUSTRATION 17 In the northeastern
United States, the annual fluctuation in Lyme
disease depends on the interaction of deer,
bacteria-carrying deer ticks, mice, oaks,
and gypsy moths.

population and can be ignored here. One interaction that is occasionally significant, however, is an indirect interaction between
white-footed mice and white oaks that is mediated by gypsy moths
(*Lymantria dispar*). These moths constitute one of the most serious
pests in northeastern broad-leafed forests, where their greedy
caterpillars feed on oak leaves and other foliage. A large population of caterpillars can completely strip oaks of their leaves, greatly
reducing acorn production or even killing the trees. The caterpillars

are not a threat every year, because gypsy moth populations are cyclical. They reach a maximum about once a decade with differences greater than a thousandfold between their highest and lowest densities. At their peak, the caterpillars can defoliate vast expanses of oak forest.

At the other end of the mouse-moth-oak chain, gypsy moths and white-footed mice are joined as prey and predator: During the summer months the mice feast on moth pupae. Because the numbers of mice and moths vary from year to year, the annual effect of mice on the moth population is not simple to forecast. When the number of mice is moderate-to-large and there are few gypsy moths, mice can consume enough pupae to hold the moth population in check. In years when the forest harbors few mice but many moths, on the other hand, the mice cannot eat enough pupae to affect the number of moths significantly. These considerations are important, because at high density the caterpillars may weaken enough oak trees to disrupt the natural cycle of mast fruiting, perhaps even forestalling a mast year. The effect of the mouse-moth-oak interaction on the tick population is difficult to evaluate, but now and then the mice check an increase in moth density, delaying the appearance of peak numbers of caterpillars that would ravage the oaks. When this happens, the mice have affected mast fruiting and the acorn production that governs the number of ticks.

Each of the individual associations between acorns and ticks is reasonable, but constructing the entire chain joining two mammals, a tick, a tree, and a moth to one another was slow work. Finally, as the ecologists verified each link, it became clear that the number of acorns on the forest floor effectively determines the future population of deer ticks. Using this conclusion and noting that 1994 was a mast year for white oaks, the ecologists predicted that 1996 would bring a large infestation of ticks in the northeastern United States. If they were right, 1996 would be a bad year for Lyme disease. In fact, the number of reported cases jumped from about 12,000 nationwide in 1995 to 16,000 in 1996, making it the worst year for

Lyme disease on record, and then fell back to about 10,000 in 1997. Most of the 1996 increase reflected more cases in the Northeast. The prediction had indeed been right.

We can now consider the place of special chemicals in the lives of these organisms. Some of these chemicals link different organisms and others act within individual species. Chemicals in the first group participate directly in the biological chain between acorns and ticks. Those in the second group contribute to a particular species' success and so influence these interactions indirectly. All the chemicals are a normal part of these creatures' lives and all have a meaningful place here. We begin with oaks and then deal with the animals.

Like many other plants, oak trees protect themselves with chemicals. Oaks synthesize defensive compounds called tannins that are in the leaves ingested by feeding caterpillars. In the caterpillar's gut the tannins bind to the proteins that are important leaf constituents, rendering them indigestible. Because these proteins are a principal source of nourishment from the leaves, the tannins substantially reduce the leaves' food value to the caterpillar. Caterpillars feeding on tannin-rich leaves grow more slowly and afford smaller pupae. The adults that emerge lay fewer eggs. The oaks' defense diminishes gypsy moth fitness, reducing the likelihood that the moths will interfere with mast fruiting. For this reason the tannins become part of our story.

Oak leaves ordinarily contain some tannins, but oak trees that survive defoliation by gypsy moths strengthen their defenses against future assaults. The new leaves they put out are richer in tannins and tougher. The trigger for this defensive response may be a chemical signal received by the trees directly from caterpillars. This is a reasonable possibility, because chemicals picked up from feeding insects' saliva induce a number of other plants to multiply their chemical defenses. Such a chemical signal passing from gypsy moth caterpillars to oak trees would also be part of our story, as it influences the oaks' defenses.

For our purposes, oaks are important because they drop the acorns that determine how many white-footed mice successfully withstand the harsh winter months. Before winter arrives, however, a multitude of other factors has influenced the number of mice available to gather the acorns. One of these factors is the mice's reproductive activity, and this happens to be under the influence of some peculiar pheromones. Because they contribute to determining the number of mice, these chemicals also enter our story. They are airborne signals present in adult-mouse urine that affect the maturation and reproductive success of young female mice.

In natural populations of white-footed mice, reproduction is restricted to a few adult females that are aided by two of these urinary signals in sustaining their privileged position. One is a remarkable pheromone that postpones the onset of puberty in young female mice. Instead of reaching sexual maturity at the normal age of forty-four days, a young female mouse exposed to adult-female urine matures about ten days late. Even bedding soiled by adult females retains enough scent to delay maturation. This postponement of puberty favors older females by reducing local competition for males as mating partners. If young females reach maturity later, their mating opportunities are reduced. However, male mice apparently prefer having more females available, not fewer, because they have their own urinary pheromone that counteracts the older females' signal. When a young female receives signals from both a male and a female, she matures normally.

Though a male mouse's presence upsets adult female control of the mating pool, this interference does not go uncontested. The females retaliate with a second extraordinary urinary signal that directly represses pregnancy in young mice. When an immature female finally reaches puberty, she soon mates with a male. However, if an older female is nearby, the young mouse has little chance of becoming pregnant. She may ovulate and mate normally, and her eggs may be properly fertilized, but her pregnancy nevertheless fails. The older female's signal somehow blocks implantation of the fertilized eggs in the young mouse's uterus, leaving her free to mate but not to bear offspring.

These pheromones increase the reproductive success of older females, but just how consequential they are for white-footed mice is not yet clear. We can reasonably assume that they affect the number and fitness of the mice in the forest and so contribute to our story. Related pheromones regulate reproduction in other species of mice, including common laboratory mice (*Mus musculus*), which have been studied in considerable detail. Other mice also have additional signals in their urine to advertise their occupancy of a particular area and to communicate their sex, sexual state, and relative age; it is likely that white-footed mice make use of such signals as well.

From mice we turn to ticks and their chemical signals. Altogether, a deer tick must obtain blood three times if it is to succeed and reproduce. At each stage, it apparently eavesdrops on host chemical signals to guide its search. On sensing odors from potential hosts, a tick climbs to the tip of a blade of grass or other vantage point where it has an opportunity of fastening on to a passing animal. There are few details established about odors attractive to larval and nymphal deer ticks, but there have been several investigations of adults.

The simplest chemical compound that attracts adult deer ticks is carbon dioxide, which is in the breath of all mammals. In one experiment designed to study this effect, investigators first dusted 120 ticks with a fluorescent powder to render these tiny creatures more visible and then released them at various distances from a carbon dioxide–baited trap. After six days, the average distance a deer tick had crawled to reach the trap was 1.8 meters (a little less than 6 feet), or 600 body-lengths for a 3-millimeter tick. This is steady progress toward the carbon dioxide source but rather slow locomotion compared with many insects and even other ticks.

All mammals exhale carbon dioxide, however, so it is not a specific attractant. Most often, adult deer ticks choose a white-tailed

deer as their host, which suggests that deer ticks are acutely sensitive to specific deer odors. Glandular secretions from the head,
hind legs, and hooves of white-tailed deer all contain mixtures of
chemicals that attract ticks. Doe urine is also effective. These observations nicely explain the frequent presence of ticks congregated
along deer trails in the woods. Responding to scents left behind by
passing deer, the ticks collect where the signals are strongest and
await potential hosts, a useful strategy since deer use the same
trails repeatedly. The chemical signals here are an indispensable
element linking deer to deer ticks. A tick unable to sense these
odors and move toward their source would have little prospect of
encountering a host.

The deer secretions that draw ticks also carry messages for
other deer, although the meaningful components may be different
for the two animals. White-tailed deer rub their foreheads on twigs
and branches, leaving behind scent marks that other deer investigate with apparent interest. The messages these marks carry are
still unclear to us, but they apparently relate to individual identity
and status. Does frequently sniff and lick at marks left by bucks,
occasionally rubbing their own foreheads over them. Because male
marks are chemically complex and their composition varies widely
from animal to animal, the marks probably distinguish one deer
from another. By secreting a somewhat different mix of the mark's
many components, each animal can possess a unique smell. The
deer in a group, then, can recognize one another individually by
odor. Concentrations of certain compounds also differ in the markings of dominant and subordinate bucks, suggesting another role of
the marks in denoting dominance.

Deer's hooves also contain scent-producing glands in small
pockets between the primary digits on all four feet. As a deer
moves about, a bit of a secretion from these glands is left behind in
its footprints. Some components differ in concentration between
dominant and subordinate animals, again pointing to communication of social position, but the secretion very likely is a trail marker
as well. When deer track one another or backtrack over their own

trails, they are presumably following this odor. Although we cannot yet interpret deer signals completely, they play a significant part in organizing and maintaining the animals' community life. As such, they must influence the number of deer introducing deer ticks into the oak forest and so become a consideration here.

Finally, we should mention the sex attractant of the gypsy moth. Along with the attractants of other economically important moths, this pheromone has received thorough investigation. It is essentially a single chemical compound known as dispalure that chemists have identified and synthesized. Receptive female moths release dispalure to attract a mate. The pheromone is carried downwind, and males react to it by flying upwind, continuing their flight toward the odor until they reach the waiting female. Without dispalure, only those male moths fortuitously close to a receptive female would find mates, and the moth reproductive rate would plummet. If the mouse-moth-oak interaction is to have its occasional effect on the tick population, dispalure too must be included in our story.

In summary, the known chemical interactions that influence the ties between acorns and ticks include: defensive compounds and, perhaps, signals between caterpillars and oaks that may affect production of both tannins and fruit; mouse reproductive pheromones that influence population size; signals attracting ticks to their hosts; deer pheromones with a variety of effects; and the gypsy-moth sex attractant. In only a few instances are the chemical identities of the active compounds known, but in many cases the operative agent is doubtless a mixture of compounds. Each of these chemicals contributes, directly or indirectly, to determining the fluctuating number of ticks in the forest. There are probably other significant chemical interactions that remain unidentified, such as the additional mouse signals mentioned above. Another possibility is tick pheromones, because a number of other ticks use several pheromones to regulate reproductive activity and aggregation of individuals into groups. Deer ticks may possess similar signals.

In another direction, changes in the forest and its inhabitants may influence the system connecting acorns and ticks, possibly establishing or removing chemical interactions. This happened after the gypsy moth was introduced into the United States from Europe around 1869 for use in silk manufacture. Until they escaped captivity in Massachusetts and invaded American forests, there were no gypsy moths available to defoliate oak trees and consequently no mouse-moth-oak connection. Mice and oaks have been in these forests for thousands of years, but their gypsy moth–mediated interaction was absent before the late nineteenth century.

Using this modest five-species system as a model, we can better picture a living world embracing all the forest's inhabitants and their many chemical interconnections. Innumerable chemical exchanges pass continually among these creatures, some having far-reaching consequences and all interrelated in unfathomed ways. The wider implications of special chemicals bring real complexity to their effects.

This complexity appeared when we looked outward from simple chemical interactions to the larger world in which they function. Another sort of complexity can be found by looking inward, by examining more closely the biological origins of a particular chemical. In our discussions we have rarely considered this inward complexity, usually ignoring where special chemicals come from, how organisms make these chemicals, or why they make one rather than another. Why do coca (*Erythroxylon coca*) and its relatives, but apparently no other plants, make cocaine? How did stick insect eggs come to offer a bit of food to ants? Why do many moth-attractant pheromones have the same general chemical structure, but trail pheromones of various ant species differ widely? We have occasionally touched on such matters, but without trying to fit special chemicals into the rest of biology. Still other areas of study, particularly genetics and the evolutionary history of organisms, open fascinating questions concerning the where, how, and why of these chemicals. Though we do not yet have detailed answers to these questions, our understanding is slowly increasing as scientists have begun to examine these issues. For the present, we shall simply illustrate inward complexity with an unusual defensive system. This is the extraordinary way a little wasp safeguards her progeny.

As we have seen, a large number of tiny parasitic wasps lay their eggs in or on the larvae of various insects. One of these wasps (*Cotesia congregata*) parasitizes large green caterpillars known as tobacco hornworms (*Manduca sexta*) and several of their close relatives. (These hornworms are the larvae of Carolina sphinx moths.) As hornworms grow, they feed insatiably. Because they particularly enjoy the leaves of tobacco, pepper, and tomato plants, they are serious agricultural pests.

In parasitizing these caterpillars, each female *Cotesia* must first locate her victim. Although hornworms can be larger than a man's little finger, they are green and easily overlooked in among a plant's leaves. The wasp has little difficulty, however, because she hunts by odor rather than sight. In response to a hornworm's depredations, a plant releases volatile defensive compounds that initially guide the wasp's search. As she nears the leaf where a hornworm is feeding, compounds in its feces steer *Cotesia* to its exact location. She lands on the caterpillar and lays her eggs, injecting as many as several hundred of them into its body through her ovipositor. As this eventually leads to the caterpillar's death, the wasps are important beneficial insects in limiting the crop damage tobacco hornworms inflict.

The larvae that hatch from *Cotesia* eggs remain inside their caterpillar host, exploiting it as a secure haven from the outer world and a convenient food supply. They feed on the caterpillar from within until they are mature and ready to pupate. Then they emerge from its body and immediately begin spinning small white cocoons, which they attach by one end to the caterpillar's back. It is not uncommon for a tobacco hornworm, still alive, to be festooned with fifty or more *Cotesia* cocoons, each resembling a diminutive grain of rice fastened to the caterpillar. Each cocoon contains

a developing pupa that should ultimately come forth as a new
adult wasp.

This arrangement is ideal for the wasps, provided the caterpillars can be prevented from destroying the wasps' offspring. Like
other creatures, tobacco hornworms have potent defenses against
foreign invasion. Unless the wasps somehow disable the hornworms' defenses, wasp eggs and larvae are doomed to a quick
death. Furthermore, wasps must disarm the hornworms without
killing them. If the caterpillars die before the larvae mature and
emerge, the larvae will die with them. For the same reason, the
wasp larvae must also avoid killing their host as they consume its
body fluids. The wasps need to keep the hornworms alive but
defenseless.

How do the wasps do this? Many kinds of parasitic wasps employ their venom to neutralize a caterpillar's defenses, adding a
dose of toxic proteins as they deposit their eggs. A female *Cotesia*
uses her venom, but for her, this is only the beginning. *Cotesia congregata* is one of several dozen kinds of wasps that inject a virus
along with their venom when laying eggs, and this virus is responsible for many of the striking effects that follow. A virus has genes
and proteins of its own and is much more complex than any single
chemical substance. With a generous stretch of definition we can
treat it as a large special chemical, providing we note one characteristic of viruses that chemical compounds lack: Given an appropriate host, a virus can replicate (reproduce) itself efficiently inside a
host cell, destroying the cell as it creates many new virus particles.

In addition to genes for its own replication, the virus that female *Cotesia* wasps deposit along with their eggs has genes for synthesizing toxic proteins that impair hornworms in several ways.
The most important of these is to disable the caterpillars' defenses
against external attack. Thirty minutes after *Cotesia* has laid her
eggs, the virus that accompanied them has spread throughout the
caterpillar's body and gained entry into its cells. Most importantly,
it has penetrated those immune cells that identify and eliminate
foreign invaders. A few hours later, these cells undergo a rapid

transformation that can readily be observed under a microscope. They begin by losing bits of their membrane and cellular contents; soon thereafter they clump together and die. These immune cells constituted the caterpillar's major defense against invasion, and now they are gone. A simple experiment demonstrates how vital these cells are. If *Cotesia* eggs taken directly from a gravid wasp and washed free of any virus adhering to them are artificially inserted into an unparasitized caterpillar, they do not survive. Normal immune cells immediately recognize them as foreign, and quickly attack and kill them. A healthy caterpillar in this case has no trouble ridding itself of wasp eggs before they hatch.

The virus's next significant assault on the hornworms arrests their normal development. Wasp larvae can grow and mature only so long as their hosts continue life as feeding caterpillars. However, left to their natural schedule, caterpillars will be ready to bury themselves in the ground and pupate before the wasp larvae are mature. For the wasps to succeed, this must not happen. The wasps must artificially extend the caterpillars' larval life and so prevent their metamorphosis. The *Cotesia* virus provides toxins that prevent pupation by interfering with the hormones that control it. Long after the wasps have emerged and long after the normal time for metamorphosis, parasitized caterpillars remain developmentally retarded. These ill-fated caterpillars will never pupate. Decorated with the wasps' tiny cocoons, they may linger as long as two weeks before dying.

The combination of the wasps' own toxins and those provided by the virus affect tobacco hornworms in other ways as well. One of these is to modify their behavior to the wasps' benefit. Parasitized caterpillars continue to feed and behave normally until about eight hours before the larvae emerge. At that time, the caterpillars cease to eat and crawl about. They show no other deficiency and their reflexes appear normal. The details of this modification are uncertain, but it appears to favor the wasp larvae. A normal, unparasitized caterpillar readily eats wasp pupae offered to it. This implies that an active caterpillar would be a threat to larvae emerging to spin their cocoons and pupate. The larvae's chance of survival is better if the caterpillar stops feeding and moving.

A natural question is how wasp and virus first became associated. Where did the wasp get such an advanced delivery system? A significant clue comes from the unexpected discovery that all the *Cotesia* virus's genes also appear in the genetic material of both male and female *Cotesia* wasps. Each wasp egg contains not only all the genes necessary to create a new wasp but also all those needed for producing the *Cotesia* virus. A wasp need never be infected with the virus, because it carries the virus's genes from birth. Not surprisingly, no individual *Cotesia congregata* has ever been found free of the virus. After initial formation by way of the wasps' genes, the virus reproduces in the female wasps' ovaries. This is not uncontrolled replication leading to cell rupture, but rather an orderly process that generates virus to be used in infecting hornworms. The wasps add a coating of this stock of virus to their eggs just before laying them.

Wasp and virus thus enjoy a peculiar and close genetic relationship. Each depends on the other for its survival, but their relationship extends even farther than this implies. Not only do wasps carry the viral genes along with their own, but the genes for the wasps' own protein toxins are very similar to the genes for some of the virus's proteins. This strongly suggests that these two sets of genes are somehow related through a common evolution. This bit of evolutionary history is still obscure, but apparently the wasps either acquired toxin genes pre-existing in the virus or else transferred some of their own toxin genes to the virus. Such movement of host genes to and from viruses is by no means uncommon, and more than one explanation for the wasp-virus relationship is conceivable. Nevertheless, one possibility now under serious consideration is quite extraordinary. This is that the *Cotesia* virus never was a free entity, but that the wasps discovered how to copy some of their own genes and package them as a novel virus for convenient transfer of toxins to hornworms. Such a discovery would be enormously advantageous to the wasps. A virus that reproduces itself and delivers toxins is much more effective than a single dose of wasp venom. It can continue to deliver more and more toxin as it spreads through the host.

Whether this remarkable explanation is correct awaits further study, as do many mysteries concerning the ability of virus and wasp to manipulate the hornworms' life processes. If the wasp has indeed created the virus for its own purposes, it will be worth considering to what extent the two can be regarded as independent entities.

Even with questions remaining to be answered, *Cotesia*'s control of its tobacco hornworm hosts can serve our initial purpose: to illustrate the inward complexity of special chemicals, to touch upon such questions as where special chemicals come from and why an organism makes one chemical rather than another. The relationship between *Cotesia congregata* and its virus affords a glimpse of this inward complexity, for in this relationship there is evidence of evolution and genetics at work, somehow linking wasp and virus. Evolution and genetics are also at work in the development of all the special chemicals we have discussed, but often in less apparent ways. Making and using a novel compound requires an organism to follow new genetic instructions. The genes carrying these new instructions ordinarily arise through changes in genes the organism already possesses, changes that come about as a part of the organism's ongoing evolution. Interpreting the origin of even the simplest special chemical demands a knowledge of a creature's evolutionary history.

Perhaps these examples of inward and outward complexity position special chemicals more realistically in the living world. Every special chemical has the possibility of impinging on the lives of many more creatures than those immediately concerned. It also has a history tied to an organism's evolution, which continues to mold the genetic directions for its use and so to control its future. Each of these interconnections leads to unanswered queries, for there is generally an unexplored narrative concerning each of these compounds. On exploration, many of these narratives will prove to

be as astonishing as the links between acorns and deer ticks, or as surprising as the unfolding account of *Cotesia congregata* and its virus. The narratives themselves then open new considerations. In this way, knowledge has grown dramatically in recent decades, yielding a detailed description of our world that was inconceivable fifty years ago. As each fresh detail augments this description, it also seems to reinforce the perception that as yet we know very little.

Capitalizing on Ecology

As you will remember, the fruits of our effort to turn chemicals to human uses include anticancer drugs, high-tech materials, and a multitude of other commodities. These products are probably the closest association most of us have with special chemicals, and so this practical topic provides a fitting place for us to end our explorations. In closing then we return to this theme of human use of natural chemicals—the extensive enterprise we previously called biotechnology. In these final pages, we want to emphasize how improved understanding of organisms' lives can point the way both to new sources of materials and to compounds more finely tuned to specific applications.

The impact of ecology on biotechnology is nicely illustrated by a new approach to antibiotics that has appeared over the past decade. For years, development of antibiotics has depended heavily on the intensive screening of microorganisms that are related to known microbial sources of these drugs. Soil bacteria are the most common source of clinically useful antibiotics, and a number of widely prescribed drugs have come from a single genus of these microbes called *Streptomyces*. There is every reason to continue the profitable screening of these bacteria, but alternative sources of drugs and alternative modes of search merit consideration as well.

For this reason, drug houses and government agencies were enthusiastic when about a decade ago, a simple realization gave rise to a novel approach to antibiotics. This was the recognition that we are not the only creatures threatened by pathogenic microbes and that some other species may well make their own antibiotics for self-protection. Of these susceptible species, those that associate in large numbers with their own kind are especially prone to microbial infection. Diseases spread readily in a crowd, and a few sick individuals can quickly unleash an epidemic. These vulnerable animals may fight infection with antibiotics that we too would find beneficial.

Animals that might be worth examining in this regard include those that form herds or flocks, as well as those that live in organized societies, such as the social insects (ants and termites, some bees and wasps). Among these prospects for study, ants offer a number of advantages. Many ant species are easily available and relatively convenient to handle in the laboratory. They have a short generation time and can undergo rapid evolutionary change. Perhaps even more pertinent is that all worker ants in a colony are normally daughters of a single queen. Their close genetic relationship facilitates the spread of infection throughout the colony and so enhances the protective advantage antibiotics would offer. Moreover, secretions from both ants and bees have antibiotic properties that were recognized decades ago. Taken together, these considerations imply that ants might employ antibiotics humans would find useful.

This prospect furnished the incentive for an investigation of disease in Australian bulldog ants (*Myrmecia gulosa*). In Australia, these ants' burning stings and painful bites are widely renowned. Ferocious, easily provoked, and 2 centimeters (more than three-quarters of an inch) long, bulldog ants are Australia's answer to the fierce army ants found in other parts of the world. Most people are content to give them wide berth.

No matter how vicious bites and stings may be, they offer no protection from pathogenic microbes. For these enemies, bulldog ants call on a different defense, as the following experiment demonstrates. Biologists produced a shallow wound in several dozen ants by pricking one leg with a fine glass tube drawn to a sharp point. Through this wound, they infected each ant with a common human intestinal bacterium (*Escherichia coli*, usually referred to as *E. coli*), and seven days later froze the infected ants and ground them into a powder. From the ant powder, they obtained two compounds that had antibiotic properties. In contrast, no antibiotic was obtained from uninfected ants otherwise treated in the same way.

When tested, the antibiotic compounds killed or inhibited the growth of two varieties of *E. coli* but had no effect on several other types of cells. These results show that in response to bacterial infection the ants elaborated an antibiotic that was selectively toxic to the pathogen. Their defense was tailored closely to their need. It is too soon to know more, but it seems that looking for new antibiotics in ants is a promising idea. Further research should establish whether ant antibiotics will lead to drugs for human use and also reveal whether other crowded species also synthesize antibiotics.

The idea of looking for antibiotics in species that associate in large numbers arose from simple considerations of ecology and natural history that pointed to these species as ones that might well have such defenses. Making connections of this sort between creatures' lifestyles and biotechnological goals has been called biorational deduction. Investigations based on this approach appeared in earlier chapters, so we have previously met the concept, if not the term. One of the studies we mentioned that was based on such a connection was the effort to design soft, tough synthetic materials modeled on the chemical structure of blue-mussel byssal threads. Another was the widespread industrial application of extremophile enzymes, substances that are useful because they function efficiently at temperatures outside the operating range of ordinary enzymes. In each case, reflection on how organisms live inspired exploration of their chemicals.

In recent years, many more biotechnological research projects have been inspired by organisms' ways of life. One is a discovery in spiders that has furnished an accessory to biomedical research. Spiders have lived for three hundred million years or more by capturing insects, paralyzing them with a neurotoxin-laden bite, and then storing them for later consumption. Spider venom paralyzes without killing, and the paralysis it induces becomes worse when the victim struggles. These intriguing properties encouraged careful study of how the venom acts. Investigation established that it interferes on the molecular level with nerve-impulse transmission. For neuroscientists this was a useful property, and subsequently spider venom became a valuable tool in their research.

A second investigation concerns the use of fungi, together with their characteristic enzymes and toxins, as selective bioherbicides. In one Australian study, a particular fungus (*Drechslera avenacea*) proved to be a virulent pathogen on wild oats (*Avena fatua*), which is a herbicide-resistant weed plant common in both Australian and North American grain fields. Because the fungus is much less toxic to wheat and barley than to the weed, it appeared to be an attractive candidate for commercialization. Research now in progress is directed toward developing it into a practical bioherbicide to control wild oats in Australian wheat fields. One issue this research must address is whether the fungus might harm other plants if it is applied widely as an agricultural herbicide. In another development in the United States, a fungus-based herbicide is already being marketed to combat northern jointvetch, a troublesome weed in rice and soybean fields. Again, the fungus ignores these crop plants while attacking the weed.

In a different arena, a new sunscreen is under development from chemical compounds produced by reef-building corals. Corals living in a reef cannot move about, and in shallow water they are continually exposed to the sun's ultraviolet rays. This radiation can be as harmful to corals as to humans, so it seemed likely that sedentary corals must somehow limit radiation damage to their bodies. Pursuing this idea in the laboratory, investigators found that corals synthesize a family of ultraviolet-absorbing compounds.

These chemicals act much as the active ingredients in our sunscreens but have larger sun-protection factors. A project is currently under way to develop and market this discovery as a powerful new sunscreen.

Another novel objective grew out of the observation that birds never feed on a particular insect, the azalea lace bug (*Stephanitis pyrioides*). These lace bugs are little creatures that specialize in sucking sap from azalea leaves, where they give rise to a spotted discoloration. A severe infestation can turn leaves white and cause them to drop off. The bugs' spiny black nymphs are gregarious and collect on the underside of leaves to feed in large groups. Although these collections of tiny bugs appear to be an easily exploitable food source, predatory birds consistently ignore them. As this behavior suggests, the nymphs are chemically protected from predation. There is now some interest in exploiting their defensive chemicals as a bird-repellent, as few chemical repellents are on the market for controlling birds that become agricultural pests. Although several repellents for mammals are available, none of them is effective against birds.

These investigations illustrate the biorational approach to desirable chemicals, but none of them demanded awareness of such complex interactions as those linking oak trees and Lyme disease. In the future, complicated situations of this sort should be better understood and make their own contribution to biotechnology. This possibility and where it may lead are admirably portrayed by a remarkable discovery involving an ant, a bacterium, and two sorts of fungi. These creatures are the principals in a second tale of ants and antibiotics.

The ants in this case are a number of related tropical species that have learned how to grow fungi. Few animals cultivate their own food, but about fifty million years ago a group of ants domesticated a fungus and began raising it for food. Since that time, the descendants of these early farmers have cultivated fungi as their dominant source of nourishment, and their entire lifestyle is built

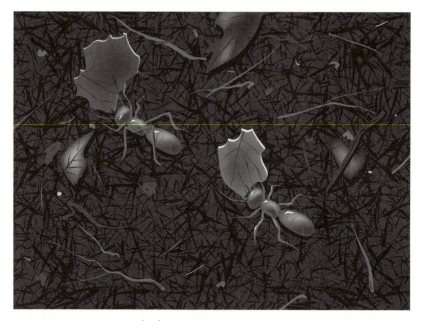

ILLUSTRATION 18 A leaf-cutting ant
makes twenty to fifty bites in cutting out a
piece of leaf to carry home and nourish the
colony's fungus gardens.

around this singular activity. Instead of collecting food for the colony, worker ants spend their days foraging to feed their fungi. The
most advanced of these ant species are leaf-cutting ants that live in
colonies consisting of thousands or even millions of individuals
and have nests that contain special chambers dedicated to fungus
gardens. These ants tirelessly strip entire trees of their leaves,
which they cut into pieces they can carry home. There they masti-
cate the leaves thoroughly, blend in some additives of their own,
and feed the pulpy mass to their fungi.

Although scientists have observed these remarkable ants at
work for decades, only in early 1999 was it recognized that the
story involves not two, but four different organisms. It appears that
fungus gardens harbor an unwanted guest. Like our own vegetable
gardens, the ants' gardens should be attractive targets for hungry,
destructive intruders. In this case, the intruder is a specialized fun-

gus (an *Escovopsis* species) that overruns the garden fungus and seriously endangers the ants' food supply.

This virulent parasite is probably an ancient pest, and the ants long ago devised an efficient response to its threat. To protect their gardens, the ants carry on their bodies antibiotic-producing bacteria that stop the parasite in its tracks. It is striking that the bacterium the ants carry is a species of *Streptomyces*, the same genus that affords many of our own antibiotics. Our development of *Streptomyces* antibiotics in the mid-twentieth century was a milestone that helped revolutionize clinical medicine, but fungus-growing ants had made the same discovery long before, probably millions of years before we existed.

These recent findings indicate that three species' lives depend on their ability to maintain two mutually beneficial arrangements. The ants must sustain and protect both their fungus and their bacterium, for without the bacterial antibiotics they could not preserve the fungus gardens that furnish their food. Similarly, without help from its ant and bacterium partners the garden fungus could not survive the parasite's invasion. At the moment, we know less about the bacterium, but in general *Streptomyces* species are highly specialized organisms. This one is associated with at least twenty-two different species of fungus-growing ants but has never been encountered elsewhere. It seems likely that it subsists only on its ant hosts, where it makes antibiotics in exchange for a secure habitat.

This complicated arrangement raises interesting questions relevant to human antibiotics. The fungal parasite has apparently failed to become resistant to *Streptomyces* antibiotics despite millennia of exposure, although pathogenic bacteria develop resistance to our antibiotics in a matter of years. Why is this so? How has the bacterial antibiotic remained effective for so long? Does it undergo structural changes to counteract resistance developing in the parasite, implying a sort of evolutionary arms race between bacterium and parasite? If so, how often is its structure modified? Also, is the antibiotic effective only against this parasitic fungus? If so, why? More generally, do any other mutually beneficial relations between organisms profit from the presence of previously unsuspected microorganisms? For instance, some termites also cultivate fungi. Do they enjoy the advantages of bacterial antibiotics as well?

Answering questions of this sort will require deep probing of complicated interactions with techniques from such specialties as genetics, biochemistry, and molecular biology. Apart from their biological importance, the answers could have far-reaching implications for new human antibiotics. A serious problem in treating infectious diseases is the resistance of an increasing number of dangerous pathogens to virtually all clinically available antibiotics. Perhaps a better understanding of how the fungus-garden drug maintains its efficacy could teach us how to circumvent resistance and design more effective antibiotics.

The mutually beneficial relationship between fungus-garden ants and their antibiotic-producing bacteria has spawned a new line of thought about antibiotics, as has the recent discovery that bulldog ants synthesize their own antibiotic compounds. Only a few years ago, no one could have foreseen these surprising links between ants and new drugs. As more ecological relationships become familiar, fresh directions for biotechnology will surely continue to appear.

We have no way of anticipating the character of these new directions or the kinds of creatures they will involve, but the diversity of currently useful organisms is truly astounding. This chapter alone has dealt with chemicals from spiders, insects, bacteria, fungi, and corals—creatures representing four major groups of organisms. We could just as readily have mentioned useful species drawn from four or five other, totally different groups. Every segment of the biological spectrum explored for useful products has made its contribution to our demand for new chemicals and materials. Perhaps the next bonanza will come from flatworms or fir trees, or perhaps from an organism yet to be discovered.

These considerations suggest that all sorts of living creatures have much to offer us from a strictly economic point of view. While there are certainly many other factors involved, these biotechnol-

ogy-related considerations constitute a strong argument for preserving the living world inviolate. Our self-interest and well-being
will be well served if we are solicitous of the living world and com-
mit ourselves to its long-term health. Besides, it is an absolutely
fascinating place.

GLOSSARY

Definitions given here reflect usage in the text. Some terms have additional or broader meanings.

algae   Marine and freshwater chlorophyll-bearing organisms that range from microscopic unicells to the giant ocean kelps 100 feet or more in length and that may be plantlike but lack such plant structures as leaves, roots, and stems.

amino acid   Any of the twenty common naturally occurring small molecules that are strung together to form peptides and proteins.

antenna   One of two moveable segmented sense organs on the head of an insect or other arthropod.

anther   The part of a flower that develops and contains pollen, usually carried on a stalk.

atom   The smallest particle of a chemical element that can exist alone.

beetle   Any member of a very large order (Coleoptera) of insects characterized by an outer pair of wings modified to form a stiff covering over the inner pair when at rest. About half of all insects and more than a quarter of *all* described species are beetles.

billion   Used in the American sense: one thousand million.

biosynthesis   Chemical synthesis carried out by a living organism.

brood   The larvae and pupae of an insect species.

catalyst   A substance that accelerates a chemical reaction without being consumed in it.

caterpillar   A moth or butterfly larva.

cellulose   A natural compound composed of many repeated sugar units that is found in plant-cell walls and in cotton and other natural fibers. It is the raw material for such manufactured goods as paper, rayon, and cellophane.

centimeter   See **units of measurement.**

chemical ecology   The scientific study of chemically mediated interactions between living organisms and their environments.

convergent evolution   The process whereby organisms living similar lives may develop similar characteristics through independent selection over evolutionary time. The wings of insects, bats, and birds, for example, are all used for flying but evolved independently of one another. Other examples appear in chapters 2 and 3.

corn   Used in the American sense: the plant (*Zea mays*) otherwise known as maize.

DNA   (Deoxyribonucleic acid) Any of various large molecules that are composed of repeated units of nucleotides and encode genetic information.

ecology   The scientific study of the interrelationship of living organisms and their environments.

enzyme   A protein molecule that acts as a biochemical catalyst.

evolution   The unifying theoretical principle of biology, based on the mutation of genes, which leads to a variation in individuals. Those individuals best suited to their environment survive to reproduce and so pass on their genes. In this way, populations evolve over time. A more extended discussion appears in chapter 1.

family   A taxonomic subdivision; see **taxonomic classification**.

fatty acid   Any of a class of acidic compounds that originate in the decomposition of natural fats.

flowering plant   A plant that reproduces sexually by way of seeds contained in an ovary. Included are most familiar plants other than ferns and conifers.

fungi   A group of about one hundred thousand nonphotosynthetic organisms that have cells with nuclei, absorb food from solution directly through their cell walls, and reproduce by way of spores. Included are molds, mushrooms, truffles, and rusts.

gene   The fundamental physical and functional unit of heredity, on a molecular level consisting of a specific portion of nucleic acid, usually DNA.

genetics   The scientific study of heredity and variation in living organisms.

genus   A taxonomic subdivision; see **taxonomic classification**.

gravid   Pregnant.

herbivore   An animal that feeds on plants.

hormone   An internal chemical signal biosynthesized at one site in an organism and transported to another, where it produces a behavioral or physiological effect.

larva   The first immature form of an insect undergoing complete metamorphosis.

lipids   A class of chemical compounds including fats, oils, and waxes.

maggot   A fly larva.

mandible   One of an insect's pair of mouth appendages that form a biting jaw.

mandibular gland   An organ situated in an insect's mandible that produces, stores, and releases a pheromone.

metamorphosis   A general term used in biology for a normal process in the life history of an organism, usually an insect, that results in a rapid, extreme change in its structure. Insects lay eggs that hatch and develop into adults through two types of metamorphosis. In incomplete metamorphosis the hatchlings look much like adults and are usually called nymphs, but they pass through several stages of growth and molting before becoming sexually mature adults. Grasshoppers and roaches provide familiar examples of this form of development. In complete metamorphosis, the hatchlings look nothing like the adults, are frequently wormlike, and are usually called larvae. After feeding and growing, a larva passes into a quiescent pupa, encased within a pupal case or cocoon. Here, a pupa undergoes structural reorganization, finally emerging from the cocoon as an adult. Bees and butterflies provide familiar examples of this form of development.

millimeter   See **units of measurement.**

mite   Any of many thousands of species of very small animals that, together with the slightly larger ticks, form a subclass of the arachnids (which also includes spiders and scorpions), which are a class of arthropods (which also includes insects and crustaceans). Ticks and mites, like spiders, differ from insects in having eight legs and no antennae. They have colonized almost all available earthly habitats and parasitized almost every organism larger than themselves.

molecule   A discrete, three-dimensional grouping of atoms connected by chemical bonds that forms the smallest indivisible unit of a chemical compound.

mollusk   A member of a large invertebrate phylum (Mollusca) that includes oysters, snails, slugs, squids, octopuses, and several less well known groups.

National Institutes of Health (NIH)   An agency of the U.S. Public Health Service responsible for developing new knowledge that will lead to better health through supporting and conducting biomedical research. Its annual budget is in excess of $15 billion.

natural history   The scientific study of species in terms of their life histories, habitat preferences, enemies, biogeography, reproduction, and related characteristics.

nematode   A roundworm and member of the invertebrate phylum Nematoda. These slender, unsegmented worms range from less than 1 millimeter to greater than 8 meters in length. Some are free living, but they have parasitized virtually every species larger than themselves.

nucleic acid   See **DNA.**

nucleotide   Any of the several small molecules that are the building blocks of nucleic acids such as DNA.

neurotoxin   A compound that interferes with the transmission of signals along a nerve fiber.

nymph   The immature form of an insect undergoing incomplete metamorphosis.

order   A taxonomic subdivision; see **taxonomic classification.**

ovary   The organ of a plant or animal that produces eggs.

ovipositor   An organ for depositing eggs.

parasite   An organism that lives with another organism (the host) in an association beneficial to the parasite and detrimental to the host. A parasite either does not kill its host or does so only over more than a single generation. Familiar examples are mistletoe and tapeworms.

parasitoid   An organism that develops in or on another organism in an association that eventually causes the death of the second organism, while the parasitoid continues its life elsewhere. Most parasitoids are wasps or flies, but several other groups are also represented.

pheromone   A chemical communication signal that functions among the members of a single species.

phylum   A taxonomic subdivision; see **taxonomic classification.**

plankton   Very small or microscopic floating or weakly swimming plant and animal life that abound in the sea, brackish water, and fresh water.

pollen   A dustlike mass of grains generated by a flowering plant that give rise to sperm cells.

predator   An organism that feeds on other organisms.

prey   An organism taken by another organism as food.

protein   A naturally occurring molecule consisting of from about fifty to several thousand amino acid units. Proteins are essential to all living organisms, serving as enzymes, structural materials, toxins, pheromones, and key components of muscle, among other functions.

protozoa   A large group of animal-like organisms made of a single cell containing one or more nuclei and ranging from several microns to several centimeters in size. Protozoa include the pathogens that cause dysentery, sleeping sickness, leishmaniasis, and malaria.

pupa   The quiescent form leading to the adult of an insect undergoing complete metamorphosis.

sex attractant   A pheromone released by one sex of a species to attract the other sex for purposes of reproduction.

species   A kind of organism, more specifically a group of interbreeding populations of an organism reproductively isolated from other such groups. See **taxonomic classification**.

stigma   The part of a flower that receives pollen grains.

synthesis   The process of building up a specific molecule from simpler ones through chemical transformations.

taxonomic classification   Biologists arrange organisms systematically into groups having similar characteristics. Every species belongs to a genus (plural *genera*), family, order, class, phylum (plural *phyla*), and kingdom, with additional levels (such as tribe and suborder) often inserted for further clarification. Similar genera form a family; similar families form an order, etc. Customary usage is to italicize the name of both genus and species but not larger groups and to capitalize the name of the genus and larger groups but not the species. For example, the common house fly, whose systematic name is *Musca domestica*, is the species *domestica* in the genus *Musca*, which also includes such closely related organisms as the face fly (*Musca autumnalis*). *Musca* and related genera form the family Muscidae; two other genera in this family are represented by stable flies and tsetse flies. The family Muscidae, in turn, is a member of the suborder Cyclorrhapha, as are the three families represented by scuttle flies, fruit flies, and flower flies, along with several others. The Cyclorrhapha are one of three suborders comprising the order Diptera, which includes the more than one hundred thousand species of two-winged "true flies," mosquitoes, gnats, and midges. The order Diptera is one of some twenty-five orders of insects (the number of orders depends on the classification scheme employed), which together comprise the class Insecta. Three other insect orders are represented by fleas, termites, and grasshoppers. The Insecta are one of several classes that make up the very large phylum Arthropoda. Two other classes of arthropods are represented by arachnids (such as spiders, ticks, and scorpions) and crustaceans (such as shrimp, crabs, and barnacles). The Arthropoda are a phylum in the kingdom Animalia, which comprises all animals. Three other animal phyla are those containing starfish, earthworms, and grizzly bears, respectively. Current classification divides living creatures into five discrete kingdoms: bacteria, fungi, plants, animals, and a kingdom comprising algae, protozoa, and a number of other less familiar groups.

tick   See **mite**.

toxin   A poison biosynthesized by a living organism.

units of measurement   Scientific measurements are expressed using an internationally agreed upon version of the metric system. The gram (28.35 grams equals 1 ounce) is the common unit of mass, the meter (1 meter equals 39.37 inches) is the common unit of length, and the liter is the common unit of volume (1 liter equals 1.057 U.S. quarts). Prefixes are attached to indicate multiples or fractions of these units. The prefixes used here include centi- (one one-hundredth), milli- (one one-thousandth), micro- (one one-millionth), and kilo- (one thousand). Thus, a centimeter is one one-hundredth of a meter and a kilogram is one thousand grams. Grains of table salt are about two hundred to five hundred micrometers in diameter and weigh one hundred to five hundred micrograms. A centimeter is about four-tenths of an inch, and a millimeter is about one twenty-fifth of an inch.

vector   An organism that transmits a pathogen.

venom   A toxin transmitted by biting or stinging.

World Health Organization (WHO)   An international body devoted to achieving the highest possible level of health for all peoples. WHO has a particularly important role in countries with governments that are unable to provide adequately for public health.

Most of these reviews and general discussions are accessible to the general reader, but a few assume a more extensive background in chemistry or biology than does the text. Many contain references to original papers. Relevant information at several levels is also available on the Internet, and the URLs (Internet addresses) of articles and pointers to this material are included.

Prologue

SLAVE-MAKING ANTS

Sudd, J. H., and N. R. Franks. *Behavioural Ecology of Ants*. Glasgow: Blackie, 1987.

Topoff, T. "Slave-making Ants." *American Scientist* 78 (1990): 520–528.

Chapter 1

CHEMICAL DEFENSE

Luszniak, M., and J. Pickett. "Self-defense for Plants." *Chemistry in Britain* 34 (July 1998): 29–32.

Rosenthal, G. A., and M. R. Berenbaum. *Herbivores: Their Interactions with Secondary Plant Metabolites*. San Diego: Academic Press, 1991.

CHEMICAL ECOLOGY

Agosta, W. *Bombardier Beetles and Fever Trees*. Reading, Mass.: Addison-Wesley, 1996.

Cardé, R. T., and W. J. Bell. *Chemical Ecology of Insects 2*. New York: Chapman and Hall, 1995.

Eisner, T., and J. Meinwald. *Chemical Ecology: The Chemistry of Biotic Inter-action*. Washington: National Academy Press, 1995.

Oldham, N. J., and W. Boland. "Chemical Ecology: Multifunctional Compounds and Multitrophic Interactions." *Naturwissenschaften* 83 (1996): 248–254.

PHEROMONES

Agosta, W. C. *Chemical Communication: The Language of Pheromones*. New York: Scientific American Library, 1992.

Chapter 2

ANT GARDENS

Davidson, D. W., and W. W. Epstein. "Epiphytic Associations with Ants." In *Vascular Plants as Epiphytes*, edited by U. Lüttge, 200. Berlin: Springer-Verlag, 1989.

ANTS

Hölldobler, B., and E. O. Wilson. *The Ants*. Cambridge: Harvard University Press, Belknap Press, 1990.

SEED DISPERSAL

Handel, S. N., and A. J. Beattie. "Seed Dispersal by Ants." *Scientific American* 263 (Aug. 1990): 76.

Chapter 3

ORCHID BEES

Dressler, R. L. "Biology of the Orchid Bees." *Annual Review of Ecology and Systematics* 13 (1982): 373–394.

POLLINATION

Barth, F. G. *Insects and Flowers*. Princeton: Princeton University Press, 1985.

Buchman, S. L., and G. P. Nabhan. *The Forgotten Pollinators*. Washington: Island Press, 1996.

Nabhan, G. P., and S. L. Buchman. "Services Provided by Pollinators." In *Nature's Services*, edited by G. C. Daily, 133–150. Washington: Island Press, 1997.

Chapter 4

LEISHMANIASIS

Peters, W., and R. Killick-Kendrick. *The Leishmaniases in Biology and Medicine*. New York: Academic Press, 1987.

TRYPANOSOMIASIS

Pepin, J., and F. Milford. "The Treatment of Human African Trypanosomiasis." *Advances in Parasitology* 33 (1994): 2–47.

YELLOW FEVER

Peterson, R. K. D. "Insects, Disease, and Military History: The Napoleonic
Campaigns and Historical Perspective." *American Entomologist*
41 (1995): 147–160.

Chapter 5

ANT-DECAPITATING FLIES

Brown, B. V. "Ant-Decapitating Flies: Nature's Executioners." *Terra* 32
(1–2) (1995): 4.

EAVESDROPPING

Stowe, M. K., T. C. J. Turlings, J. H. Loughrin, W. J. Lewis, and J. H. Tumlinson. "The Chemistry of Eavesdropping, Alarm, and Deceit." *Pro-
ceedings of the National Academy of Sciences, USA* 92 (1995): 23–28.

FLOWER MITES

Colwell, R. K. "Stowaways on the Hummingbird Express." *Natural History*
94(7) (1985): 56–63.

PARASITIC WASPS

Tumlinson, J. H., W. J. Lewis, and L. E. Vet. "How Parasitic Wasps Find
Their Hosts." *Scientific American* (Mar. 1993): 100–106.

Chapter 6

CATERPILLARS AND ANTS

DeVries, P. J. "Singing Caterpillars, Ants, and Symbiosis." *Scientific American* 267 (Oct. 1992): 76–82.

MIMICRY

Stowe, M. K. "Chemical Mimicry." In *Chemical Mediation of Coevolution*, edited by K. Spencer, 513–580. San Diego: Academic Press, 1988.

SENITA CACTUS AND *DROSOPHILA*

Kircher, H. W., and W. B. Heed. "Phytochemistry and Host Plant Specificity in *Drosophila*." *Recent Advances in Phytochemistry* 3 (1970): 191–209.

Chapter 7

BACTERIAL PHEROMONES

Losick, R., and D. Kaiser. "Why and How Bacteria Communicate." *Scientific American* 276 (Feb. 1997): 68–73.

DRAGON FISH PIGMENTS

Zurer, P. "Dragon Fish Uses Chlorophyll Derivatives to See in the Dark." *Chemical and Engineering News* (8 June 1998): 10.

Chapter 8

EXTREMOPHILES

Gross, M. *Life on the Edge: Amazing Creatures Thriving in Extreme Environments.* New York: Plenum Press, 1998.

Madigan, M. T., and B. L. Marrs. "Extremophiles." *Scientific American* 279 (Apr. 1997): 82–87.

TYRIAN PURPLE

Clark, R. J. H., C. J. Cooksey, M. A. M. Daniels, and R. Withnall. "Indigo, Woad, and Tyrian Purple: Important Vat Dyes from Antiquity to the Present." *Endeavour* 17 (1993): 191–199.

Chapter 9

DRUGS FROM MARINE SOURCES

Faulkner, J. "Chemical Riches from the Deep." *Chemistry in Britain* 31 (1995): 681–684.

DRUG TESTING

Stevenson, R. "Gold Standard for Drugs." *Chemistry in Britain* 34 (1998) (9): 31–35.

PSORALENS

Pathak, M. A., and T. B. Fitzpatrick. "The Evolution of Photochemotherapy with Psoralens and UVA (PUVA): 200 B. C. to 1992 A. D." *Journal of Photochemistry and Photobiology B: Biology* 14 (1992): 3–22.

TRADITIONAL MEDICINE

Balick, M. J., and P. A. Cox. *Plants, People, and Culture: The Science of Ethnobotany.* New York: Scientific American Library, 1996.

Chapter 10

YUCCAS AND THEIR MOTHS

Ramsay, M., and J. R. Schrock. "The Yucca Plant and the Yucca Moth." *Kansas School Naturalist*, 41 (1994–95) (2): 1–9.

Chapter 11

COTESIA WASPS AND THEIR VIRUSES

Beckage, N. E. "The Parasitic Wasp's Secret Weapon." *Scientific American* 279 (Nov. 1997): 82–87.

Moore, Janice. "Parasites that Change the Behavior of Their Hosts." *Scientific American* 250 (May 1984): 100–108.

ECOLOGY
Muir, J. *My First Summer in the Sierra*. Cambridge, Mass.: The Riverside Press, 1911.

LYME DISEASE
Ostfield, R. S., C. G. Jones, and J. O. Wolff. "Of Mice and Mast." *Bioscience* 46 (1996): 323.

Internet Addresses
All addresses are preceded by ⟨http://⟩.

CHEMICAL ECOLOGY
International Society of Chemical Ecology: ⟨www.isce.ucr.edu/Society/⟩

ECOLOGY
Ecology pointers: ⟨ecoregion.ucr.edu/lee/Ecology-List.html⟩

FLOWERING PLANTS
Internet Directory for Botany: ⟨www.ou.edu/cas/botany-micro/idb/ botvasc.html⟩

GENERAL
Scientific American (magazine): ⟨www.sciam.com/⟩
The Scientist (magazine): ⟨www.the-scientist.lib.upenn.edu/⟩

INSECTS
The Wonderful World of Insects: ⟨www.earthlife.net/insects/six.html⟩

LYME DISEASE
Lyme disease pointers: ⟨www.bitgroup.com/informedpatient/bluecross/ search2.html⟩

PARASITES
Parasitology pointer: ⟨dspace.dial.pipex.com/town/plaza/aan18/ urls.htm⟩

INDEX

DATE DUE
